EPITOME
OF
COPERNICAN
ASTRONOMY
&
HARMONIES
OF THE WORLD

EPITOME
OF
COPERNICAN ASTRONOMY

&

HARMONIES
OF THE WORLD

JOHANNES KEPLER

GREAT MINDS SERIES

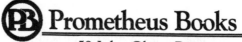 **Prometheus Books**

59 John Glenn Drive
Amherst, NewYork 14228-2197

Published 1995 by Prometheus Books

59 John Glenn Drive, Amherst, New York 14228–2197,
716–691–0133. FAX: 716–691–0137.

Library of Congress Cataloging-in-Publication Data

Kepler, Johannes, 1571–1630.
   [Epitome astronomiae copernicanae. English]
   Epitome of Copernican astronomy ; and, harmonies of the
world / Johannes Kepler ; translated by Charles Glenn Wallis.
       p.   cm. — (Great minds series)
   Originally published: Annapolis : The St. John's Bookstore,
c1939.
   ISBN 1–57392–036–3 (pbk. : alk. paper)
   1. Astronomy—Early works to 1800. 2. Kepler, Johannes,
1571–1630. Harmonices mundi. English. I. Kepler, Johannes,
1571–1630. Harmonices mundi. English. II. Title. III. Title:
Harmonies of the world. IV. Series.
QB41.K3413   1995
520—dc20   95–35655
                                          CIP

Printed in the United States of America on acid-free paper.

Titles on Science in
Prometheus's Great Minds Series

See the back of this volume for a complete list of titles in Prometheus's
Great Books in Philosophy and Great Minds series.

JOHANNES KEPLER, the great German astronomer, was born at Weil, in the duchy of Wurttemberg, on December 27, 1571, into a family of obscure and impoverished nobility. Kepler began school at Leonberg in 1577, but family bankruptcy soon forced the child to leave school and go to work as a field hand. Since Kepler's mental aptitude, along with his bodily infirmity (he had suffered from smallpox as a child), seemed to suit him for a theological vocation, he was sent as a charity student to the seminary at Adelberg in 1584 and later to Maulbronn. His academic brilliance assured him entrance to the University of Tübingen (1588), where, in addition to the classical curriculum, he studied astronomy under Michael Mästlin.

In 1594 Kepler gave up a career in the ministry to accept the chair of astronomy at Graz. While at Graz, Kepler discovered an imaginary relation between the six then known planets and the five regular solids of geometry—the tetrahedron, the cube, the octahedron, the dodecahedron, and the icosahedron. This discovery, published in his *Precursor of Cosmographic Dissertations, or the Cosmographic Mystery (Prodromus dissertationum cosmographicarum seu mysterium cosmographicum)* (1596), brought Kepler his first fame and put him in contact with Galileo and the Danish astronomer Tycho Brahe.

In 1598 Ferdinand, the Catholic archduke of Styria, issued an edict of banishment against Protestant preachers and professors, forcing Kepler to flee with his wife and family to the Hungarian border. Though reinstated in his post at Graz, in 1600 Kepler accepted an offer from Tycho Brahe to become his assistant at his observatory near Prague. Following Brahe's death one year later, the emperor Rudolph II appointed Kepler imperial mathematician. Combining Brahe's records with his own researches, Kepler published a series of works which made him famous through Europe. First he wrote, to satisfy the emperor's astrological proclivities, *On the More Certain Foundations of Astrology* (*De fundamentis astrologiae certioribus*) (1602); on the strength of this work, Kepler's "prognostics" were in high demand. The *Optical Part of Astronomy* (*Astronomiae pars optica*) (1604), completed as the *Dioptrics* (1611), contained important discoveries in the theory of vision, and a notable approximation toward the true law of refraction. But Kepler's greatest achievement during this period was an elaboration of a new theory of the planets titled the *New Aetiological Astronomy, or Celestial Physics together with Commentaries on the Movements of the Planet Mars* (*Astronomia nova aetiologetos, seu physica coelestis tradita commentariis de motibus stellae Martis*) (1609): in this seminal work, two of the cardinal principles of modern astronomy were worked out: the laws of elliptical orbits and of equal areas.

In 1611 Kepler's first wife died of typhus and his favorite child of smallpox; that same year Rudolph was deposed by his brother Matthias, and Kepler's salary was in arrears. While still in the position of court astronomer, he accepted an offer to become mathematician to Upper Austria, moving to Linz in 1613. Happy and productive once more in a second marriage, Kepler resumed his astronomical observations.

The dozen years Kepler spent at Linz saw the publication of his most important astronomical works: *The Harmonies of the World* (*Harmonices mundi*) (1619) set forth an elaborate system of celestial harmonies depending on the various and varying velocities of the several planets; the *Epitome of Copernican Astronomy* (*Epitome astronomiae Copernicanae*) (1618–21), in dialogue form, was a textbook of Copernican science, remarkable for the prominence given to "physical astronomy" and for the extension to the Jovian system of the laws recently discovered to regulate planetary motion. This latter work utilized a portion of the material Kepler had been gathering for a project to comprehend the whole scheme of the heavens to be called the *Hipparchus*. Kepler also published the *Rudolphine Tables* (1627), the compilation of his own and Brahe's astronomical tables.

Kepler's last years were dogged by continuing financial troubles. Still owed twelve thousand florins by the imperial treasury, in 1628 Kepler entered the service of Duke Wallenstein of Friedland, who assumed responsibility for the debt. However, in lieu of the back pay due him, the duke offered Kepler a professorship at Rostock, an offer Kepler declined. In 1630 he traveled to Ratisbon to plead his case before the Diet. Shortly after his arrival, however, Kepler fell ill and died on November 15, 1630.

Other works by Kepler include *De stella nova* (1696), on a brilliant star that appeared suddenly on September 30, 1604, and remained visible for seventeen months; *Nova stereometria doliorum* (1613), which helped lead the way toward the discovery of calculus; and *Somnium seu astronomia lunari* (posthumously published 1634), a scientific satire.

Epitome of Copernican Astronomy

Contents

3

Contents

For the information of the reader, the translator here appends a table of contents of the first and untranslated part of the work:

BOOK I. Concerning the Principles of Astronomy in General and Those of the Doctrine on the Sphere Specifically.

Part I. Concerning the shape of the Earth, its magnitude, and the method of measuring it.

Part II. Concerning the shape of the heavens.

Part III. Concerning the nature and altitude of the air, which envelopes the land and ocean and its distinction from the ether, which is diffused throughout all the heavens.

Part IV. Conerning the place of the Earth in the world and its proportionality in relation to the world.

Part V. Concerning the daily movement of the Earth.

BOOK II. Concerning the Sphere and its Circles.

BOOK III. Concerning the Doctrine of the First Movement—called the Doctrine on the Sphere.

Part I. Concerning the rising and setting of the stars.

Part II. Concerning the ascensions and disclination of the signs or points on the ecliptic.

Part III. Concerning the year and its parts, and concerning the days and their increases or decreases in length.

Part IV. Concerning the seasons of the year and the magnitudes of the zones.

Part V. On the apparition and occultation of the stars at different times of the year.

BOOK FOUR

Herein the natural and archetypal causes of Celestial Physics *that is, of all the magnitudes, movements, and proportions in the heavens are explained and thus the Principles of the Doctrine on the Schemata are demonstrated.*

This book is designed to serve as a supplement to Aristotle's On the Heavens.

TO THE READER

IT has been ten years since I published my *Commentaries on the Movements of the Planet Mars.* As only a few copies of the book were printed, and as it had so to speak hidden the teaching about celestial causes in thickets of calculations and the rest of the astronomical apparatus, and since the more delicate readers were frightened away by the price of the book too; it seemed to my friends that I should be doing right and fulfilling my responsibilities, if I should write an epitome, wherein a summary of both the physical and astronomical teaching concerning the heavens would be set forth in plain and simple speech and with the boredom of the demonstrations alleviated. I did that before many years had passed. But meanwhile various delays came between the book and publication: the little book itself was not up to date in spots, and, unless I am mistaken, it was also incomplete in the form in which it was given, and even the plan of publication began to totter. For in the "doctrine concerning the sphere"—published before three years were up—I seemed to certain people to be more diffuse in arguing about the diurnal movement or repose of the earth than befitted the form of an epitome. Accordingly I reflected that if the readers had not digested that part, which was however absent from no epitome of astronomy, all the more strange to them would be this Fourth Book, which airs so many new and unthought-of things concerning the whole nature of the heavens—so that you might doubt whether you were doing a part of physics or astronomy, unless you recognized that speculative astronomy is one whole part of physics.

On the other hand, I considered that this was a matter for the sake of my amplifying which and impressing it upon the public, *i.e.*, for the sake of my writing this little book, many men of letters had become my friends: that these speculations could not be omitted, unless I spent my devotion in giving attention to the darkness of a doctrine of schemata which was robbed of its proper principles. At least, necessity—how I wish she were sometimes less importunate!—cut short this disputation: for necessity makes that which cannot be done otherwise seem to be undertaken as if by design. The press

5

groaned and the work on the doctrine of schemata was being struck off, when its lawful godfather, whom I mentioned in the foreword to the Spherical Doctrine, attained his former state, and was sleeping, or perhaps giving up the ghost, and as the liberality of this most eminent patron was paying for the parts of this book, it became necessary for me suddenly to set out and to break off the work. At that same time the printers had reached the end of the Fourth Book and the Frankfort market-day was at hand. I decided that it would be best if the Fourth Book, the subject-matter of which includes both physics and astronomy, were also published separately; whence, according to the choice of the astronomer buying, it could be passed over, or inserted into the rest of the epitome. Kind reader, you have the reasons for this publication, and I hope you will find them satisfactory.

But as regards this branch of philosophizing: it will not be out of keeping with the job at hand if I here set down in advance some things from the recent letter which I wrote to a man who is intimate with a great Prince and is himself also a great man. In this letter a comparison was undertaken between this book—or the related work *On the Harmonies*, published in the previous year—and Aristotle's books *On the Heavens* and *Metaphysics*; and this philosophy [*i.e.*, modern astronomy] was cleared of the worn-out charges of being esoteric and seeking after novelty.

Accordingly, these are the excerpts from the aforesaid letter which have to do with the present undertaking:

It seems to me I have nothing to worry about in the case of Aristotle: His Most Serene Highness is a Platonist in philosophy and a Christian in religion: His Most Serene Highness cannot dislike whatever is the more convincing, whether it be that the world was first made at a fixed beginning in time as was my work *On thë Harmonies*, or will be destroyed at some time, or is merely liable to destruction, like the alterations of the ether and the celestial atmosphere; nor will he ever prefer the Master Aristotle to the truth of which Aristotle was ignorant.

But if His Most Serene Highness has a high opinion of Aristotle, wheresoever he reveals the mysteries of philosophy, if he makes any serious remark or any praiseworthy attempt; for indeed he is the man who in *On the Heavens* (Book II, Chapter 5) asks: "For what reason are there many movements?" So I ask: "What are the reasons for the number of the planets?" He asks in the following chapter: "For what reason are the heavens borne from east to west rather than from west to east?" So I ask: "Why is any planet moved with so much speed, no more, no less?" In Chapter 9 he asks: "Do the stars give forth sounds which are modulated [*contemperatos*] harmonically?" and answers no: I split up his judgment, for I grant that no sounds are given forth but I affirm and demonstrate that the movements are modulated according to harmonic proportions. In Chapter 10 he asks "about the order of the spheres, the intervals, and the ratio of the movements to the orbital circles"; but he merely asks and fails in the attempt. Not only do I answer these questions with most luminous demonstrations by means of the five regular solids, but also I add the number of the planets, which has been deduced from the Archetype, so that it may be clear that the world is created. In Chapter 12 he asks: "Why in the descent from the upper to the lower planets are not the movements of the single planets found to be more manifold?" and he pronounces a judgment most elegantly tempered by the modesty of confession and the wisdom of assertion. "Let us try," he says, "to say only that which appears as true; for we judge that the readiness" even to put forward what is probable "is worthy of being characterized as modesty rather than presumption, if anyone, in things concerning which there are very great difficulties, is content—in order to satisfy his thirst for philosophy—with even slight discussions such as these." But I myself, led on by this same praiseworthy thirst for philosophy, first wiped away from the eyes of astronomy those mists of the multiplicity of movements in the single planets: then I gave a demonstration of the following: that the movement of the planet is not uniform throughout its whole circuit—as Aristotle argued in Chapters 6 and 7; but that in reality the movement is increased and decreased at places in its period which are fixed and are opposite to one another; and I explained the efficient or intsrumental causes of this increase as the lessening of the interval between the planet and the sun, from which as from a source that movement arises. Then, as in each and every planet there is a very fast movement and a very slow movement and in a fixed proportion, I did not merely raise the question as to the reason for this proportion in the single planets separately and in all the planets in relation to one another; and why Saturn and Jupiter have middling eccentricities, Mars a great eccentricity, the Sun and Venus slight eccentricities, and Mercury a very great eccentricity; but I also brought forward a solution of this very great difficulty, and not a trifling discussion but one wholly legitimate; and I took my solution from the Archetype of the harmonic cosmos: whence it is established that this cosmos cannot be better than it is and that it is impossible that the world should not have been created at a fixed beginning in time.

This attempt of mine ought not to have been checked by shyness, but should have been brought forth into the light with strength of mind, namely, with the highest confidence in the visible works of God—if one has leisure for knowledge of them—or at the exhortation of Aristotle himself, who judged that in these questions you should not suppress or be silent about probabilities any more than about fully explored certainties. Then he is that same Aristotle who, in the *Metaphysics*, Book XII, Chapter 8, in which place he built up the most sublime part of his philosophy, the part concerning the gods and the number of them; who, I say, sends his students to the astronomers and who defers to the astronomers in respect to their authority and the weight of their testimony; indeed he would never have scorned Tycho Brahe or even myself, if that fatal necessity of the generations had made us contemporaries. For he orders his students "to read through both," that is to say, Eudoxus and Callippus, for the one had corrected the errors of the other; and today that would be to read both Ptolemy and Tycho: "but to follow" not, he says, the more ancient, but "the more accurate." And so, if Aristotle is dear to that most just Prince, I call Aristotle to witness that he has suffered no injury, if the astronomer, using the arguments which modern times have put forward concerning the heavens, has indicated that creatures arose in the heavens and will disappear once more—in opposition to the opinion of him who alleges experience, but experience not sufficiently long.

As regards the academies, they are established in order to regulate the studies of the pupils and are concerned not to have the program of teaching change very often: in such places, because it is a question of the progress of the students, it frequently happens that the things which have to be chosen are not those which are most true but those which are most easy. And by that division in things which makes different people form different judgements, it so happens that certain people are in error contrary to their own opinion. It seems to me that the truth concerning the mutable nature of the heavens can be taught conveniently; but someone else judges that students and teachers equally are thrown into confusion by this doctrine. But it is not without its use in explaining even those parts of the philosophy of Aristotle which are clearly false, as Book VIII of the *Physics* concerning celestial movement and Book II of *On the Heavens* concerning the eternity of the heavens—so that a comparison could be made between the philosophy of the gentiles and the truth of Christian dogma. Accordingly, if certain subtleties which are difficult to grasp should not be laid before beginners, or if they should not be preferred to the accepted and necessary teachings, it does not follow that therefore those things should neither be written nor read privately. You can count few academies in which it is a part of the program to explain the *Metaphysics* of Aristotle: yet Aristotle wrote the *Metaphysics* too, a very useful work in the judgement of the professors on all the faculties. Therefore, in order that no one should consider His Most Serene Highness blameworthy, if he observes the rules of the academies, and if he believes that the honour of the academics—even if they have sinned greatly in judgement—should be defended against presumptuous critics, against untimely quarrellers: so in turn I do not let myself be easily persuaded that this most wise Prince will seek to have all people remain publicly and privately inside the boundaries of academic philosophy; and to have no one labour privately in bringing forward these things, that is to say, in the manifestation of the works of God.

But His Most Supreme Highness will not pick a fight concerning the heavens; for he knows that the philosophers speak of the visible heavens; and Christ of the invisible heavens, or, as the schools say, of the empyrean, or, as the simple Christians take it, of the blessed seats, which no corruption will ever touch: since not Tycho, not I, but Christ Himself pronounces concerning this visible world: "Heaven and Earth shall pass away," and the Psalmist, "they shall grow old like a garment"; and Peter, "They shall be destroyed root and all, and be consumed by burning in the fire." And that will occur in order that the alterations in the heavens should not destroy their eternity, if there should be such an eternity, just as the terrestrial alterations, which are perrennial and return in a circle, destroy the Earth's eternity which was equally believed by Aristotle. But this kind of argument against Aristotle will perhaps seem too contentious. Therefore let us use his own testimony instead; for he is not everywhere consistent: in the *Metaphysics* he attributes movement to the celestial bodies for its own sake and teaches "that they are moved in order that they may be moved"; but in *On the Heavens*, being admonished by the things themselves, he attributes something or other like the terrestrial, something multiplex and turbulent to the stars or rather to their movers, who by means of these mechanisms and movements seek another end outside of the movement itself, and one mover attains this end with more difficulty than another: in this way, as a matter of fact, he adduces the fewness of movements in the moon as witness of the inferior condition of the moon and its closer kinship to the Earth. For he means to say that the celestial bodies which cannot wholly attain the highest end by their own nature do not employ many motions; and that it would have been wholly useless for the Earth to have a movement to attain that end, but that the Earth is absolutely at rest there; that the moon progresses somewhere and stretches out towards that end; the the higher bodies attain the end, but by many movements; and the highest heaven, by the one simple movement. And so he compares the actions, the πράξεις of the moon—that is the word he uses—to the uniform life of plants, but the πράξεις of the higher bodies, to the more varied life of animals. Yet he makes all those bodies to be in need of these actions because they have their end and their blessedness outside of themselves. Accordingly, in the epilogue to the Fifth Book of the *Harmonies*, I wish for Aristotle as my reader and critic; as it is not right that I should wish to take up any more of the time of His Most Serene Highness, the highest judgement of the Prince. I am sure of one thing at least, that if he would direct the cultivated power of his mind toward those things which Aristotle wrote and toward my epilogue, everything would be agreed between us, and he would by his own judgement harmonize the discord which now, as you predict, he might feel between us.

In order to counter the envious charge of novelty-hunting, it would be first in my program, even though His Most Serene Highness can easily see all things for himself, to warn him fully of the distinction between the love—or thirst, to use the Aristotelian word—for the knowledge of natural things and the lust for contradicting and holding the opposite opinion. All philosophers, whether Greek or Latin, and all the poets too, recognize a divine ravishment in investigating the works of God: and not merely in investigating them privately but even in teaching them publicly: and it can be inferred that the false charge of esoteric novelty-hunting cannot cling to this ravishment.

There is God in us, and our warmth comes from His move-
ments: This Spirit has descended from the heavenly seats.

There is no need of this declamation before you, or before His Most Supreme Highness: only I must make some further mention of the boundary posts. For the boundary posts of investigation should not be set up in the narrow minds of a few men. "The world is a petty thing, unless everyone finds the whole world in that which he is seeking," as Seneca says. But the boundary posts of true speculation are the same as those of the fabric of the world; but the Christian religion has put up some fences around false speculation which is on the wrong track, in order that error may not rush headlong but may become in other respects harmless in itself. Antiquity teaches us by examples how vainly man sets up boundary posts where God has not set them up: how severely all the astronomers were blamed by the first Christians. Did not Eusebius write of an astronomer that he preferred to desert Christianity—I suppose because he was excommunicated—rather than his profession? Who today would opine that Eusebius is to be imitated? Did not those who taught that there were antipodes seem to Tertullian and to Augustine to be over-wise? And, indeed, there was a Virgil Bishop of Salisbury who was removed from his office because he dared to assert this same fact. How many times were the Roman philosophers exiled from the city? And at that, under the ancient manners, wherewith the Roman State was established. Yet today we set up academies everywhere: we order that philosophy be taught, that astronomy be taught, that the antipodes be taught.

But I even in private free myself from the blame of seeking after novelty by suitable proofs: let my doctrines say whether there is love of truth in me or love of glory: for most of the ones I hold have been taken from other writers: I build my whole astronomy upon Copernicus' hypotheses concerning the world, upon the observations of Tycho Brahe, and lastly upon the Englishman, William Gilbert's philosophy of magnetism. If I rejoiced in novelty, I could have devised something like the Fracastorian or Patrician systems. Just as one who rejoices in occupations but rarely in companions, never of himself descends to dice or to a game of chess; similarly for me there is so much importance in the true doctrine of others or even in correcting the doctrines which are not in every respect well established, that my mind is never at leisure for the game of inventing new doctrines that are contrary to the true. Whatever I profess outwardly, that I believe inwardly: nothing is a worse cross for me than—I do not say, to speak what is contrary to my thought—to be unable to utter my inmost sentiments. I know that many innovators are produced by the same affect; but they are easily argued out of the error which seduces them. No one shows that I have committed an error. But because certain people cannot grasp the subtleties of things, they lay the charge of novelty-hunting upon me.

I now descend to the work itself, the *Harmonies:* I do not doubt that he who condemns the itch to devise new things and the presumption to profess new and grandiose things will find in the epilogue to the Fifth Book[1] that which he will mark critically. For here the sun-spots and little flames are brought forward as evidence of there being exhalations from the sun which are analogous to exhalations from the Earth: here things corresponding to the generation of animals are established as occurring in the planets—here the confines of the mysteries of Christian religion are touched: we knock at the doors of the science of the Magi, of theurgy, of the idolatry of the Persians, and of those who worship the sun as god—as the interjection of frequent warnings does not dissimulate.

Accordingly, if what has been said so far concerning these esoteric things is not satisfactory: at any rate let this be impressed upon His Most Serene Highness: that this chapter contributes nothing in its own right except conjectures; and although it adds a good deal to the form of the work: because—as the opening of the chapter has it—reason itself leads "from the Muses to Apollo": nevertheless, since the other parts of the work are established by means of their proper demonstrations, the chapter, or epilogue, can be considered as cut off from the rest. For even without the epilogue, the following thesis is upheld by incontrovertible demonstrations: *that in the farthest movements of any two planets, the universe was stamped with the adornment of harmonic proportions; and, accordingly, in order that this adornment might be brought into concord with the movements, the eccentricities which fell to the lot of each planet had to be brought into concord.* The most wise Prince will easily reckon how great an addition this makes in illustrating the glory of the fabric of the world, and of God the Architect.

But if, however, even this inquiry is accused of being esoteric: I indeed confess that the head of astronomy is struck off. And since astronomy is studied either for its own sake as a philosophy or for the sake of making astronomical

[1]*See Harmonies of the World*, pp. 240-245.

predictions; then, if I am to cast my ballot in the question of future contingencies, His Most Serene Highness repudiates any secondary end for this exact and subtle investigation of physical causes which does not offer itself for the uses of daily life: therefore the taking away from me of the primary end slays this whole subtle astronomy and plainly makes it useless.

Nevertheless, in order that I may arm myself against this eventuality also: I will grant that this work of mine, the *Harmonies*, is nothing except as it were a certain picture of the edifice of astronomy; and though it may be erased at the pleasure of him who spits upon it, nevertheless the house called astronomy stands by itself: and I know that astronomy is not condemned by His Most Serene Highness but is held of great value on account of its certitude in predicting movements: perhaps, therefore, he will judge its architect—who is almost the only renovator after the Master Tycho and who thought it worth while to devote his life to this work—to be not unworthy of his favour.

These extracts from the letter, most of which have to do with the investigation of very hidden causes which is to be viewed in this little book, should be spoken and understood. And now it is time for the reader to pass on to the little book.

FIRST BOOK ON THE DOCTRINE OF THE SCHEMATA

On the Position, Order, and Movement of the Parts of the World; or, on the System of the World

[433] *What is the subject of the doctrine of the schemata?*

The proper movements of the planets; we call them the secondary movements; and the planets, the secondary movables.

Why do you call them the proper movements of the planets?

1. Because the apparent daily movement—with which the doctrine on the sphere is concerned—and which is common to both the planets and the fixed stars, and so to the whole world, is seen to travel from the east to the west; but the far slower single movements of the single planets travel in the opposite direction from west to east; and therefore it is certain that these movements cannot depend upon that common movement of the world—which we have discussed so far—but should be assigned to the planets themselves, and thus they are generically proper to the planets.

2. But even if in these proper movements of the single [434] planets from west to east there is also present something common, not diurnal but annual, which is extrinsic and betrays that its cause lies in eyesight alone, outside the truth of the thing; and which meanwhile makes the planet in its proper movement have the appearance of retrograding, that is, from east to west, nevertheless because this common movement is so woven into the single periods of the single planets, and so variously transformed, that at first glance you cannot discern what is common to all the planets and what is proper to each: accordingly this whole composite movement of each planet, as it meets the eyes, is said to be proper to each planet specifically; especially since this movement which is common to many does not have its origin in that first

common movement of the whole world, but in the proper movement of each planet.

How many parts are there to the doctrine of the schemata?

Above (in Book I, page 15), the whole doctrine was divided into its three proper parts: the first, concerning the principles wherewith Copernicus demonstrates the secondary movements—the material of Book IV; the second, concerning the machinery whereby these movements are laid before the eyes, *viz.*, concerning the eccentric and similar circles—the material of Book V; and the third, concerning the apparent movements of the single planets and the common accidents of the planets taken together—the material of Book VI; and the fourth part, which is common to the doctrines on the sphere and on the schemata, concerns the apparent movement of the eighth sphere—the material of Book VII.

What are the hypotheses or principles wherewith Copernican astronomy saves the appearances in the proper movements of the planets?

They are principally: (1) that the sun is located at the centre of the sphere of the fixed stars—or approximately at the centre—and is immovable in place; (2) that the single planets move really around the sun in their single systems, which are compounded of many perfect circles [435] revolved in an absolutely uniform movement; (3) that the Earth is one of the planets, so that by its mean annual movement around the sun it describes its orbital circle between the orbital circles of Mars and of Venus; (4) that the ratio of its orbital circle to the diameter of the sphere of the fixed stars is imperceptible to sense and therefore, as it were, exceeds measurements; (5) that the sphere of the moon is arranged around the Earth as its centre, so that the annual movement around the sun—and so the movement from place to place—is common to the whole sphere of the moon and to the Earth.

Do you judge that these principles should be held to in this Epitome?

Since astronomy has two ends, to save the appearances and to contemplate the true form of the edifice of the world—of which I have treated in Book I, folia 4 and 5—there is no need of all these principles in order to attain the first end: but some can be changed and others can be omitted; however, the second principle must necessarily be corrected: and even though most of these principles are necessary for the second end, nevertheless they are not yet sufficient.

Which of these principles can be changed or omitted and the appearances still be saved?

Tycho Brahe demonstrates the appearances with the first and third principles changed: for he, like the ancients, places the Earth immobile, at the centre of the world; but the sun—which even for him is the centre of the orbital circles of the five planets—and the system of all the spheres he makes to go around the Earth in the common annual movement, while at the same time in this common system any planet completes its proper movements. Moreover, he omits the fourth principle altogether and exhibits the sphere of the fixed stars as not much greater than the sphere of Saturn.

[436] *What in turn do you substitute for the second principle and what else do you add to the true form of the dwelling of the world or to what belongs to the nature of the heavens?*

Even though the true movements are to be left singly to the single planets, nevertheless these movements do not move by themselves nor by the revolutions of spheres—for there are no solid spheres—but the sun in the centre of the world, revolving around the centre of its body and around its axis, by this revolution becomes the cause of the single planets going around.

Further, even though the planets are really eccentric to the centre of the sun: nevertheless there are no other smaller circles called epicycles, which by their revolution vary the intervals between the planet and the sun; but the bodies themselves of the planets, by an inborn force [*vi insite*], furnish the occasion for this variation.

What, then, will the material of Book IV be?

Book IV will contain celestial physics itself, or the form and proportions of the fabric of the world and the true causes of the movements. This will be the primary function of the astronomer—as we said in Book I, folium 5, namely, the demonstration of his hypotheses.

Review the principal parts of Book IV.

There will be three principal parts of Book IV.

The first is on the bodies themselves; the second, on the movements of those bodies; the third, on the real accidents of the movements.

For the first part will teach the conformation of the whole universe, its division into parts or principal regions; the place of the sun at its centre; the number, magnitude, and order or position of the planetary spheres; and lastly, the ratios of all the bodies of the world to one another.

The second part will teach the revolution of the sun around its axis, and its effect in making the planets revolve; the causes of the proportionality of the movements among themselves, *i.e.*, of the periodic [437] times; the immobility of the centre of the sun and the annual movement of the centre of the Earth around the sun; the revolution of the Earth around its axis and its effect in making the moon revolve; the additional help in moving the moon given by the light of the sun; and what the causes of the proportions between the day, month, and year are.

The third part will disclose the causes of the threefold irregularity of the altitude, longitude, and latitude in the single planets—and how these irregularities are doubled in the moon by the force of the illumination from the sun.

PART I

1. ON THE PRINCIPAL PARTS OF THE WORLD

[438] *What do you judge to be the lay-out of the principal parts of the world?*

The Philosophy of Copernicus reckons up the principal parts of the world by dividing the figure of the world into regions. For in the sphere, which is the image of God the Creator and the Archetype of the world—as was proved in Book I—there are three regions, symbols of the three persons of the Holy

Trinity—the centre, a symbol of the Father; the surface, of the Son; and the intermediate space, of the Holy Ghost. So, too, just as many principal parts of the world have been made—the different parts in the different regions of the sphere: the sun in the centre, the sphere of the fixed stars on the surface, and lastly the planetary system in the region intermediate between the sun and the fixed stars.

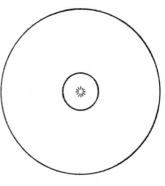

I thought the principal parts of the world are reckoned to be the heavens and the earth?

Of course, our uncultivated eyesight from the Earth cannot show us any other more notable parts—as was said in Book I, folia 8, 9, 10—since we tread upon the one with our feet and are roofed over by the other, and since both parts seem to be commingled and cemented together in the common limbo of the horizon—like a globe in which stars, clouds, birds, man, and the various kinds of terrestrial animals are enclosed.

But we are practised in the discipline which discloses the causes of things, shakes off the deceptions of eyesight, and carries the mind higher and farther, outside of the boundaries of eyesight. Hence it should not be surprising to anyone that eyesight should learn from reason, that the pupil should learn something new from his master which he did not know before—namely, that the Earth, considered alone and by itself, should not be reckoned among the primary parts of the great world but should be added to one of the primary parts, *i.e.*, to the planetary region, the movable world, and that the Earth has the proportionality of a beginning in that part; and that the sun in turn should be separated from the number of stars and set up as one of the principal parts of the whole universe. But I am speaking now of the Earth in so far as it is a part of the edifice of the world, and not of the dignity of the governing creatures which inhabit it.

By what properties do you distinguish these members of the great world from one another?

The perfection of the world consists in light, heat, movement, and the harmony of movements. These are analogous to the faculties of the soul: light, to the sensitive; heat, to the vital and the natural; movement, to the animal; harmony, to the rational. And indeed the adornment [*ornatus*] of the world consists in light; its life and growth, in heat; and, so to speak, its action, in movement; and its contemplation—wherein Aristotle places blessedness—in harmonies. Now since three things necessarily come together for every affection, namely, the cause *a qua*, the subject *in quo*, and the form *sub qua*—therefore, in respect to all the aforesaid affections of the world, the sun exercises the function of the efficient cause; the region of the fixed stars that of the thing forming, containing, and terminating; and the intermediate space, that of the subject—in accordance with the nature of each affection. Accordingly, in all these ways the sun is the principal body of the whole world.

For as regards light: since the sun is very beautiful with light and is as if the eye of the world, like a source of light or very brilliant torch, the sun illuminates, paints, and adorns the bodies of the rest of the world; the intermediate space is not itself light-giving, but light-filled and transparent and the channel through which light is conducted from its source, and there exist in this region the globes and the creatures upon which the light of the sun is poured and which make use of this light. The sphere of the fixed stars plays the role of the river-bed in which this river of light runs, and is as it were an opaque and illuminated wall, reflecting and doubling the light of the sun: you have very properly likened it to a lantern, which shuts out the winds.

Thus in animals the cerebrum, the seat of the sensitive faculty imparts to the whole animal all its senses, and by the act of common sense causes the presence of all those senses as if arousing them and ordering them to keep watch. And in another way, in this simile, the sun is the image of common sense; the globes in the intermediate space of [440] the sense-organs; and the sphere of the fixed stars of the sensible objects.

As regards heat: the sun is the fireplace [*focus*] of the world; the globes in the intermediate space warm themselves at this fireplace, and the sphere of the fixed stars keeps the heat from flowing out, like a wall of the world, or a skin or garment—to use the metaphor of the Psalm of David. The sun is fire, as the Pythagoreans said, or a red-hot stone or mass, as Democritus said—and the sphere of the fixed stars is ice, or a crystalline sphere, comparatively speaking. But if there is a certain vegetative faculty not only in terrestrial creatures but also in the whole ether throughout the universal amplitude of the world—and both the manifest energy of the sun in warming and physical considerations concerning the origin of comets lead us to draw this inference—it is believable that this faculty is rooted in the sun as in the heart of the world, and that thence by the oarage of light and heat it spreads out into this most wide space of the world—in the way that in animals the seat of heat and of the vital faculty is in the heart and the seat of the vegetative faculty in the liver, whence these faculties by the intermingling of the spirits spread out into the remaining members of the body. The sphere of the fixed stars, situated diametrically opposite on every side, helps this vegetative faculty by concentrating heat, as they say; as it were a kind of skin of the world.

As regards movement: the sun is the first cause of the movement of the planets and the first mover of the universe, even by reason of its own body. In the intermediate space the movables, *i.e.*, the globes of the planets, are laid out. The region of the fixed stars supplies the movables with a place and a base upon which the movables are, as it were, supported; and movement is understood as taking place relative to its absolute immobility. So in animals the cerebellum is the seat of the motor faculty, and the body and its members are that which is moved. The Earth is the base of an animal body; the body, the base of the arm or head, and the arm, the base of the finger. And the movement of each part takes place upon this base as upon something immovable.

Finally, as regards the harmony of the movements: the sun occupies that place in which alone the movements of the planets [441] give the appearance of magnitudes harmonically proportioned [*contemperatarum*]. The planets themselves, moving in the intermediate space, exhibit the subject or terms, wherein the harmonies are found; the sphere of the fixed stars, or the circle of the zodiac, exhibits the measures whereby the magnitude of the apparent movements is known. So too in man there is the intellect, which abstracts universals and forms numbers and proportions, as things which are not outside of intellect; but individuals [*individua*], received inwardly through the senses are the foundation of universals; and indivisible [*individuae*] and discrete unities, of numbers; and real terms of proportions. Finally, memory, divided as it were into compartments of quantities and times, like the sphere of the fixed stars, is the storehouse and repository of sensations. And further, there is never judgment of sensations except in the cerebrum; and the effect of joy never arises from a sense-perception except in the heart.

Accordingly, the aforesaid vegetating corresponds to the nutritive faculty of animals and plants; heating corresponds to the vital faculty; movement, to the animal faculty; light, to the sensitive; and harmony, to the rational. Wherefore most rightly is the sun held to be the heart of the world and the seat of reason and life, and the principal one among three primary members of the world; and these praises are true in the philosophic sense, since the poets honour the sun as the king of the stars, but the Sidonians, Chaldees, and Persians—by an idiom of language observed in German too—as the queen of the heavens, and the Platonists, as the king of intellectual fire.

These three members of the world do not seem to correspond with sufficient neatness to the three regions of a sphere: for the centre is a point, but the sun is a body; and the outer surface is understood to be continuous, yet the region of fixed stars does not shine as a totality, but is everywhere sown with shining points discrete from one another; and finally, the intermediate part in a sphere fills the whole expanse, but in the world the space between the sun [442] *and the fixed stars is not seen to be set in motion as a whole.*

As a matter of fact, the question indicates the neatest answer concerning the three parts of the world. For since a point could not be clothed or expressed except by some body—and thus the body which is in the centre would fail of the indivisibility of the centre—it was proper that the sphere of the fixed stars should fail of the continuity of a spherical surface, and should burst open in the very minute points of the innumerable fixed stars; and that finally the middle space should not be wholly occupied by movement and the other affections, nor be completely transparent, but slightly more dense, since it could not be altogether empty but had to be filled by some body.

Are there solid spheres [orbes] *whereon the planets are carried? And are there empty spaces between the spheres?*

Tycho Brahe disproved the solidity of the spheres by three reasons: the first from the movement of comets; the second from the fact that light is not refracted; the third from the ratio of the spheres.

For if spheres were solid, the comets would not be seen to cross from one sphere into another, for they would be prevented by the solidity; but they cross from one sphere into another, as Brahe shows.

From light thus: since the spheres are eccentric, and since the Earth and its surface—where the eye is—are not situated at the center of each sphere; therefore if the spheres were solid, that is to say far more dense than that very limpid ether, then the rays of the stars would be refracted before they reached our air, as optics teaches; and so the planet would appear irregularly and in places far different from those which could be predicted by the astronomer.

The third reason comes from the principles of Brahe himself; for they bear witness, as do the Copernican, that Mars is sometimes nearer the Earth than the sun is. But Brahe could not believe this interchange to be possible [443] if the spheres were solid, since the sphere of Mars would have to intersect the sphere of the sun.

Then what is there in the planetary regions besides the planets?

Nothing except the ether which is common to the spheres and to the intervals: it is very limpid and yields to the movable bodies no less readily than it yields to the lights of the sun and stars, so that the lights can come down to us.

If it is ether, then it will be a material body having density. Therefore will not its matter resist the movable bodies somewhat?

On the contrary, the ether is more rarefied than our air, since it is very pure, being spread over a space which is practically immense.

How do you prove this?

In optics, by refractions. For our air, which is contiguous to the ether, causes a refraction of approximately 30'. But water contiguous to air causes a refraction of approximately 48°, whence the ratio of the density of water to air, and of air to ether is somehow established by taking the cubes of the numbers. For 30' is contained approximately 100 times in 48°; and in squares, that is 10,000 times, and in cubes 1,000,000 times. Therefore air is that many times more rarefied than water, and ether than air.

Nevertheless the matter of the ether is not absolutely null: are the stars therefore still impeded by it?

We can without any inconvenience grant such a small impediment of movement and such a small resistance of the ether to the movable bodies, just as even before this it must be granted that they offer some resistance on account of the proper matter of their bodies, as will be made clear below. And what if no resistance should be granted to the ether, [444] since it is fairly credible that the ether which surrounds the movable globe the most closely accompanies the globe on account of the very great limpidity [of the ether]?

2.   On the Place of the Sun at the Centre of the World

By what arguments do you affirm that the sun is situated at the centre of the world?

The very ancient Pythagoreans and the Italian philosophers supply us with some of those arguments in Aristotle (*On the Heavens*, Book, II, Chapter 13); and these arguments are drawn from the dignity of the sun and that of the place, and from the sun's office of vivification and illumination in the world.

State the first argument from dignity.

This is the reasoning of the Pythagoreans according to Aristotle: the more worthy place is due to the most worthy and most precious body. Now the sun— for which they used the word "fire," as sects purposely hiding their teachings— is worthier than the Earth and is the most worthy and most precious body in the whole world, as was shown a little before. But the surface and centre, or midpoint, are the two extremities of a sphere. Therefore one of these places is due to the sun. But not the surface; for that which is the principal body in the whole world should watch over all the bodies; but the centre is suited for this function, and so they used to call it the Watchtower of Jupiter. And so it is not proper that the Earth should be in the middle. For this place belongs to the sun, while the Earth is borne around the centre of its yearly movement.

What answer does Aristotle make to this argument?

1. He says that they assume something which is not granted, namely, that the centre [445] of magnitude, *i.e.*, of the sphere, and the centre of the things, *i.e.*, of the body of the world, and so of nature, *i.e.*, of informing or vivifying, are the same. But just as in animals the centre of vivification and the centre of the body are not the same—for the heart is inside but is not equally distant from the surface—we should think in the same way about the heavens, and we should not fear for the safety of the whole universe or place a guard at the centre; rather, we should ask what sort of body the heart of the world of the centre or vivification is and in what place in the world it is situated.

2. He tries to show the dissimilarity between the midpart of the nature and the midpart of place. For the midpart of nature, or the most worthy and precious body, has the proportionality of a beginning. But in the midpart of place is the last, in quantity considered metaphysically, rather than the first or the beginning. For that which is the midpart of quantity, *i.e.*, is the farthest in, is bounded or circumscribed. But the limits are that which bounds or circumscribes. Now that which goes around on the outside, and limits and encloses, is of greater excellence and worth than that which is on the inside and is bounded: for matter is among those things which are bounded, limited, and contained; but form, or the essence of any creature, is of the number of those things which limit, circumscribe, and comprehend. He thinks that he has proved in this way that not so much the midpart of the world as the extremity belongs to the sun, or as he understood it, to the fire of the Pythagoreans.

How do you rebut this refutation of Aristotle's?

1. Even if it be true that not in all creatures and least in animals is the principal part of the whole creature at the centre of the whole mass: however, since we are arguing about the world, nothing is more probable than this. For the figure of the world is spherical, and that of an animal is not. For animals need organs extending outside themselves, with which they stand upon the ground, and upon which they may move, and with which they may take within themselves the food, drink, [446] forms of things, and sounds received from outside. The world on the contrary, is alone, having nothing outside, resting on itself immobile as a whole; and it alone is all things. And so there is no reason why the heart of the world should be elsewhere than in the centre in order that what it is, *viz.*, the heart, might be equally distant from all the farthest parts of the world, that is to say, by an interval everywhere equal.

2. Furthermore, as regards his telling us to ask what sort of body the principal part of the whole universe is: he is confused by that riddle of the Pythagoreans and believes that they claim that this element is principal. He is not wrong however in telling us to do that. And accordingly we, following the advice of Aristotle, have picked out the sun; and neither the Pythagoreans in their mystical sense nor Aristotle himself are against us. And when we ask in what place

in the world the sun is situated, Copernicus, as being skilled in the knowledge of the heavens, shows us that the sun is in the midpart. The others who exhibit its place as elsewhere are not forced to do this by astronomical arguments but by certain others of a metaphysical character drawn from the consideration of the Earth and its place. Both we and they set a value upon these arguments; and they themselves too by means of these arguments do not show but seek the place of the sun. So if when seeking the place of the sun in the world, we find that it is the centre of the world; we are doing just as Aristotle; and his refutation does not apply to us.

3. As regards the fact that Aristotle, directly contradicting the Pythagoreans ascribes vileness to the centre, he does that contrary to the nature of figures and contrary to their geometrical or metaphysical consideration.

For above in Book i, the centre was absolutely not last in the sphere, but wholly its most regular beginning of generation in the mind, and it manifests the likeness of the Holy Trinity, in shadowing forth God the Father, who is the First Person.

4. Finally it can be seen by anyone that he who judges as a physicist of those things which are geometrical does not do rightly, unless what he questions concerning matter and form [447] had been taken over by analogy from a consideration of geometrical figures. For indeed, in solid quantities the inward corporeality, everywhere spread out equally and not by itself partaking of any figure, is a true image of matter in physical things; but the outward figure of the corporeality, composed of fixed surfaces which bound the solidity, represent the

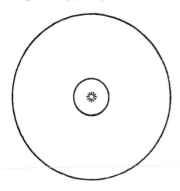

form in physical things. And so this comparison is permitted to him simply; but it appears from that that he plays equivocally with "midpart" [*medium*]. For though the Pythagoreans spoke of the inmost point of the sphere [as the midpart or centre]; he understood the whole space within the surface as comprehended by the word "midpart." Accordingly we must grant him the victory as regards the space, but it is a useless victory; for the Pythagoreans and Copernicus win as regards the midpart of all this space. For even if the midpart as a space does not deserve the name of limit; nevertheless as centre it does deserve this name. And in this respect [448] it must be added to the forms and boundaries: since above (in Book i) the centre was the origin of generation of the sphere, metaphysically considered.

Prove by means of the office of the sun that the centre is due to it.

That has already been partly done in rebutting the Aristotelian refutation. For (1) if the whole world, which is spherical, is equally in need of the light of the sun and its heat, then it would be best for the sun to be at the midpart, whence light and heat may be distributed to all the regions of the world. And that takes place more uniformly and rightly, with the sun resting at the centre than with the sun moving around the centre. For if the sun approached certain regions for the sake of warming them, it would draw away from the opposite regions and would cause alternations while it itself remained perfectly simple. And it is surprising that some people use jokingly the similitude of light at the centre of the lamp, as it is a very apt similitude, least fitted to satirize this opinion but suited rather to painting the power of this argument.

(2) But a special argument is woven together concerning light, which presupposes fitness, not necessity. Imagine the sphere of the fixed stars as a concave mirror: you know that the eye placed at the centre of such a mirror gazes upon itself everywhere: and if there is a light at the centre, it is everywhere reflected at right angles from the concave surface and the reflected rays come together again at the centre. And in fact that can occur at no other point in the concave mirror except at the centre. Therefore, since the sun is the source of light and eye of the world, the centre is due to it in order that the sun—as the Father in the divine symbolizing—may contemplate itself in the whole concave surface—which is the symbol of God the Son—and take pleasure in the image of itself, and illuminate itself by shining and inflame itself by warming. These melodious little verses apply to the sun:

> *Thou who dost gaze at thy face*
> *and dost everywhere leap back*
> *from the navel of the upper air*
> *O gushing up of the gleams flowing*
> *through the glass emptiness, Sun,*
> *who dost again swallow thy reflections.*

Nevertheless Copernicus did not place the sun exactly at the center of the world?

It was the intention of Copernicus to show that this node common to all the planetary systems—of which node we shall speak below—is as far distant from the centre of the sun as the ancients made the eccentricity of the sun to be. He established this node as the centre of the world, and was compelled to do so by no astronomical demonstration but on account of fitness alone, in order that this node and, as it were, the common centre of the mobile spheres would not differ from the very centre of the world. But if anyone else, in applying this same fitness, wished to contend that we should rather fear to make the sun differ from the centre of the world, and that it was sufficient that this node of the region of the moving planets should be situated very near, even if not exactly at the centre—anyone who wished to make this contention, I say, would have raised no disturbance in Copernican astronomy. So, firstly, the last arguments concerning the place of the sun at the centre are nevertheless unaffected by this opinion of Copernicus concerning the distance of this node from the sun. But secondly we must not agree to the opinion of Copernicus that this node is distant from the centre of the sun. For the common node of the region of the mobile planets is in the sun, as will be proved below; and so by some probable arguments either the one or the other point is set down at the centre of the sphere of the fixed stars, and by the same arguments the other point is brought to the same place, even with the approval of Copernicus.

3. ON THE ORDER OF THE MOVABLE SPHERES

How are the planets divided among themselves?

Into the primary and the secondary. The primary planets are those whose bodies are borne around the sun, as will be shown below; the secondary planets

are those whose own circles are arranged not around the sun but around one of
the primary planets and who also share in the movement of the primary planet
around the sun. Saturn is believed to have two such secondary planets and to
draw them around with itself: they come into sight now and then with the help
of a telescope. Jupiter has four such planets around itself: *D, E, F, H.* The Earth
(*B*) has one (*C*) called the moon. It is not yet clear in the case of Mars, Venus,
and Mercury whether they too have such a companion or satellite.

Then how many planets are to be considered in the doctrine on schemata?

No more than seven: the six so-called primary planets: (1) Saturn, (2)
Jupiter, (3) Mars, (4) the Earth—the sun to eyesight, (5) Venus, (6) Mercury,

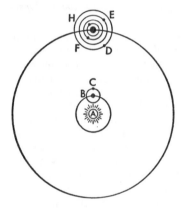

and (7) only one of the secondary planets, the
moon, because it alone revolves around our
home, the Earth; the other secondary planets
do not concern us who inhabit the Earth, [451]
and we cannot behold them without excellent telescopes.

*In what order are the planets laid out: are they
in the same heaven or in different heavens?*

Eyesight places them all in that farthest
and highest sphere of the fixed stars and
opines that they move among the fixed stars.
But reason persuades men of all times and
of all sects that the case is different. For if the
centres of all the planets were in the same
sphere and since we see that to sight they
are fairly often in conjunction with one an-
other: accordingly one planet would impede the other, and their movements
could not be regular and perpetual.

But the reasoning of Copernicus and ancient Aristarchus, which relies upon
observations, proves that the regions of the single planets are separated by very
great intervals from one another and from the fixed stars.

*What is the difference here between the reasoning of Copernicus and that of the
ancients?*

1. The reasoning of the ancients is merely probable, but the demonstration
of Copernicus, arising from his principles, brings necessity.

2. They teach only that there is not more than one planet in any one sphere:
Copernicus further adds how great a distance any planet must necessarily
be above another.

3. Now the ancients built up one heaven upon another, like layers in a wall,
or, to use a closer analogy, like onion skins: the inner supports the outer;
for they thought that all intervals had to be filled by spheres and that the
higher sphere must be set down as being only as great as the lower sphere of a
known magnitude allows; and that is only a material conformation. Copernicus,
having measured by his observation the intervals between the single spheres,
showed that there is such a great distance between two planetary spheres,
that it is unbelievable that it should be filled with spheres. And so this lay-out

of his urges the speculative mind to spurn matter and the contiguity of spheres and to look towards the investigation of the formal lay-out or archetype, with reference to which the intervals were made.

4. The ancients, with their material structure, were forced to make the planetary or mobile world many parts greater than Copernicus was forced to do with his formal lay-out. But Copernicus, on the contrary, made the region of the mobile planets not very large, while he made the motionless sphere of the fixed stars immense. The ancients do not make it much greater than the sphere of Saturn.

5. The ancients do not explain and confirm as they desire the reason for their lay-out; Copernicus establishes his lay-out excellently by reasons.

What do you mean by the reasons for the lay-out of the spheres, and how is Copernicus outstanding in this respect?

Aristotle teaches in *On the Heavens* (Book II, Chapter 10) that nothing is more consonant with reason than that the times of revolution of each planet should correspond to the altitude or amplitude of its sphere. Now for the ancients, the highest planet was the same as the slowest, namely, Saturn, because it takes 30 years. Jupiter follows it in place and in time, and takes 2 years; Mars, which takes less than 2 years, follows Jupiter. But for the ancients, this proportionality was changed in the remaining planets. For unless you grant to the Earth an annual movement around the sun, then the sun, Venus, and Mercury—three distinct planets—have the same time of revolution of a year, nevertheless they give them different spheres: the upper to the sun, the middle to Venus, and the third to Mercury. Finally they give the lowest place to the moon, as it takes the shortest time, namely a month.

But Copernicus, postulating that the Earth moves around the sun, keeps the same proportion of movement and time in all the planets. For him the sun

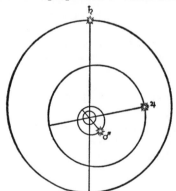

Schema of Saturn, Jupiter, Mars, and the Earth

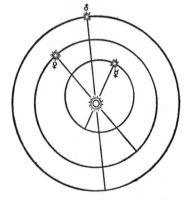

[454] Schema of the Earth, Venus, and Mercury, with orbit of the Earth enlarged

is at the centre of the world and is thus the farthest in; it is without the revolution of the centre, that is to say, it is motionless with respect to the centre and the axis. But a few years after this, the body [453] of the sun was perceived to move around its motionless axis more quickly than the space of one month. Mercury, the nearest, circles around the sun in the smallest sphere and completes its revolution in 3 months; around this sphere moves Venus in a larger sphere and in a longer period of time, *viz.*, 7½ months. Around the heaven of Venus moves the Earth with its satellite the moon—for the moon is a secondary planet, whose proportionality is not counted among the primary planets—and it revolves in a period of 12 months. After, follow Mars, Jupiter, and Saturn, as with the ancients, each with its satellite. After Saturn, comes the sphere of the fixed stars—and it is distant by such an immense interval that it is absolutely at rest.

What measure does Copernicus use in measuring the intervals of the single planets?

We must use a measure so proportioned that the other spheres can be compared, a measure very closely related to us and thus somehow known to us: such is the amplitude of the sphere whereon the centre of the Earth and the little sphere of the moon revolve—or its semidiameter, the distance of the Earth from the sun. This distance, like a measuring rod, is suitable for the business. For the Earth is our home; and from it we measure the distances of the heavens; and it occupies the middle position among the planets and for many reasons—on which below—it obtains the proportionality of a beginning among them. [455] But the sun, by the evidence and judgment of our sight, is the principal planet. But by the vote of reason cast above, the sun is the heart of the region of moving planets proposed for measurement. And so our measuring rod has two very signal termini, the Earth and the sun.

How great therefore are the intervals between the single spheres?

The Copernican demonstrations show that the distance of Saturn is a little less than ten times the Earth's from the sun; that of Jupiter, five times; that of Mars, one and one-half times; that of Venus, three-quarters; and that of Mercury, approximately one-third.

And so the diameter of the sphere of Saturn is less than twice the length of its neighbour Jupiter's; the diameter of Jupiter is three times that of the lower planet Mars; the diameter of Mars is one and one-half times that of the terrestrial sphere placed around the sun; the diameter of the Earth's sphere is more than one and one-third that of Venus; and that of Venus is approximately five-thirds or eight-fifths that of Mercury. However, it should be noted that the ratios of the distances are different in other parts of the orbits, especially in the case of Mars and Mercury.

What is the cause of the planetary intervals upon which the times of the periods follow?

The archetypal cause of the intervals is the same as that of the number of the primary planets, being six.

I implore you, you do not hope to be able to give the reasons for the number of the planets, do you?

This worry has been resolved, with the help of God, not badly. Geometrical reasons are co-eternal with God—and in them there is first the difference between the curved and the straight line. Above (in Book i) it was said that the curved somehow bears a likeness to God; the straight line represents creatures. And first in the adornment of the world, the farthest region of the fixed stars has been made spherical, in that geometrical likeness of God, because as a corporeal God—worshipped by the gentiles under the name of Jupiter—it had to contain all the remaining things in itself. Accordingly, rectilinear [456] magnitudes pertained to the inmost contents of the farthest sphere; and the first and most beautiful magnitudes to the primary contents. But among rectilinear magnitudes the first, the most perfect, the most beautiful, and most simple are those which are called the five regular solids. [457] More than 2,000 years ago Pythagoreans said that these five were the figures of the world, as they believed that the four elements and the heavens—the fifth essence—were conformed to the archetype for these five figures.

But the truer reason for these figures including one another mutually is in order that these five figures may conform to the intervals of the spheres. Therefore, if there are five spherical intervals, it is necessary that there be six spheres: just as with four linear intervals, there must necessarily be five digits.

What are these five regular figures?
The cube, tetrahedron, dodecahedron, icosahedron, and octahedron.

How are these figures divided, and into what classes?
The cube, tetrahedron, and dodecahedron are primary; the octahedron and the icosahedron are secondary.

Why do you make the former primary and the latter secondary?
The three former figures have a prior origin, and the most simple angle (*i.e.*, trilinear), and their own proper planes. The two latter have their origin in the primary figures, and a more composite angle made from many lines, and borrowed planes.

What is the order of the primary figures?
They are said to be primary merely with respect to the secondary; but even among themselves they have this order of priority: cube, tetrahedron, dodecahedron. For in those figures there appears the first of all metaphysical oppositions, that between the same and the other, or the different. Sameness is seen in the cube, and difference in the remaining two figures; and between these figures there is also the first geometrical contrariety, namely, the contrariety between the greater-than and less-than. For the cube is [458] the thing itself [*res ipsa*], the tetrahedron is less than the cube, and the dodecahedron is greater than the cube; or, the cube is the first solid figure gener-

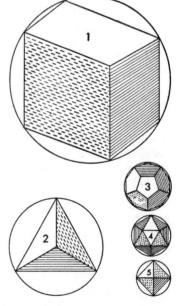

ated, the tetrahedron is the first of the solid figures cut out of the cube, and the dodecahedron is the first of the composite figures made by adding to, and covering over, the cube. This same idea is dominant in their planes: the tetragon, the triangle, and the pentagon. For the tetragon is generated first of all by drawing the most simple and regular lines, as was said in Book I; and it is broken up into two triangles; but the pentagon is composed of three suitable triangles.

Explain the generation, primacy, and form of the cube.

Rectilinear magnitudes have an origin visible to the mind; the spherical, as was said above, brings a certain character of eternity or of eternal generation. For with a sphere postulated, the point at its centre is postulated, and so are the infinite points on its surface. Therefore, line arises from the flowing of point to point; surface arises from the sideways flowing of the line; and body, from the sideways flowing of the surface. If the flowing of the point is straight and also the shortest, there arises a straight line bounded by two points. If the flowing of the straight line is such that all its points flow equally, a parallelogram arises bounded by four lines; and if the parallelogram flows in the same way, the parallelepiped arises, bounded by six planes. Again, if the flowing of the line is equal to the flowing straight line, and the line along which the flowing takes place makes any angle with the flowing line except a right angle, there arises the plane called the rhomboid, whose sides are equal. But if the line makes a right angle, it is a square which arises. And if the square also flows, there arises the cube, the six planes of which are all squares and are thus equal to one another. Now the shortest is prior to the crooked; and the equal and similar is prior to the unequal and dissimilar, and the straight, or right, to the oblique. Therefore, in this way among the lines generated, the straight line is prior—for the circle is posterior to the plane, and the plane to the straight line; and among surfaces the square is prior. Thus among magnitudes, that which exists perfectly [459] (*i.e.*, with three dimensions), that is to say, the cube, is shown to be first among bodies.

Explain the primacy of the tetrahedron among the segments, and the mode of section of the cube, and its form.

By subtracting from the bodies, so that something lesser exists, the other solid

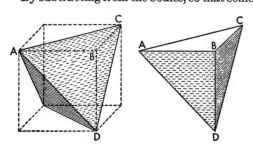

figures usually come to be. We must judge that the first of them is the solid figure which exists if the first figure generated (*viz.*, the cube) is cut most simply and most equally. But the section—among the sections which designate a new plane figure—is not equal or more simple than when you cut through four angles of the cube completely. For you are cutting out the same number of equilateral tetrahedrons—and the single tetrahedrons *A*, *C*, and *D* have a solid right angle *B* above the triangular base. There remains as it were the bowels of the cube, namely, the fifth tetrahedron, similar to itself in all respects, and bounded by four equilateral triangles. But if you make the section of the cube of which I have spoken in Book I, there will not be five but six irregular tetrahedrons. So the tetrahedron is the first figure coming from the decreasing of the bodies. But it is a third part of the body of the cube cut, and any angle cut away, such as *BACD*, is the sixth part of the same whole.

[460] *Explain the origin of the dodecahedron by addition, and give the reasons for its posteriority among the three primary bodies, and its priority among the bodies generated by addition.*

As in subtracting from the cube, four planes are constructed in place of the four angles of the cube which have been cut off; four angles remain to the tetrahedron, but the angles are decreased and are still of the same species (*i.e.*, trilinear), so too, if we wish to construct the first of the increased bodies, or of those which are greater than the cube, in place of the planes of the cube we construct angles, but we transmit the angles of the cube as clothed and increased, though they also remain trilinear; or, what leads to the same thing, upon the twelve sides of the cube the same numbers of planes are to be built, just as, in the former case, upon the six sides of the tetrahedron the same number of squares were built. For as the cube roofs over the tetrahedron, so this increased solid figure which we are investigating roofs over the cube.

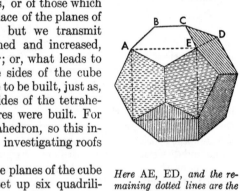

[461] But if instead of the single planes of the cube we set up single angles—if we set up six quadrilinear angles, because the six planes of the cube are quadrilaterals—the eight trilinear angles of the cube remain. Therefore the figure would be mixed. Therefore in order that the trilinear angle may remain in the augmented figure, two angles are to be constructed upon the

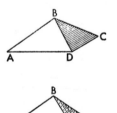

single planes of the cube—and not one angle alone; that is to say, six prisms, such as the former one *BCAED*, not six pyramids, such as the one here, *BADC*. And in order that there may always be one plane common to two contiguous prisms, let this plane be constructed across one side of the cube. And these six prisms are slightly less than the cube upon which they are placed. And so by the augmentation twelve angles are made—and by the addition of the eight angles of the cube, the sum of the angles is twenty.

Here AE, ED, *and the remaining dotted lines are the sides of the roofed-over cube.* AED *is the plane of the cube; and the two angles* B *and* C *come to be instead of it; and* A *and* E *the angles of the cube remain. But upon the side of the cube* AE, *the pentagon* ABCE *is constructed, as upon the side* ED, *the pentagon* ECD.

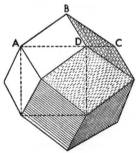

How do you infer the form of the plane of the dodecahedron?

The angles of the figure, as was said already, should be twenty; and each single angle is bounded by three lines, any one of which adjoins two angles; and so there are twenty times three or sixty points. But two points determine one line. Therefore there are thirty lines or sides to the figure, and there are potentially sixty with respect to the planes of the figure. For any [462] side of the figure adjoins two planes. But the division of the sixty lines or plane sides by the twelve planes—which are necessary for this solid figure—gives a quotient of five. Therefore the planes are quinquelateral. So among the augmented figures the dodecahedron, having pentagonal planes, is again first.

What is the origin of the secondary figures and why are there only two?

Three other figures correspond to the cube, the tetrahedron, and the dodecahedron; but one of them coincides with its primary figure. And these secondary figures are generated by subtracting from the three primary figures, but by subtraction of a different kind, where a line is not left in the place of the plane, but an angle, *i.e.*, in the place of the surface of the primary figure, there is—not a line of the secondary figure, but—a point; while the number of lines remains. But at the same time—as before—a plane of the secondary figure is generated in place of the angle of the primary figure. And the plane is triangular, because the angle of the primary figure is trilinear, and by joining together the centres of three planes of the primary figure a solid angle is constructed. So these figures are generated secondly as if the bowels of the first figures.

For whatever appears outwardly falls away from the cube, and there remain of the cube only six centres, as it were the navels of six planes. And there are six angles to the new figure. And [463] because the cube has eight angles, the figure gets eight plane equilateral triangles in their place. Hence it is called an octahedron; and it is the sixth part of its cube.

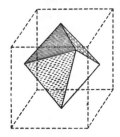

As regards the tetrahedron: in place of its plane triangles four angles are constructed; and in place of its four angles four triangles are constructed; and a figure arises which is the same as its primary figure. And so it is not judged to be new. However, it is the twenty-seventh part of the tetrahedron in which it is inscribed.

As regards the dodecahedron: [464] in place of the twelve bases it gives the twelve angles of the new figure; and in place of its twenty angles it gives twenty triangular bases: whence the figure is called an icosahedron. And it is slightly less than half the size of its original dodecahedron.

One of the primary figures was produced by subtracting from the cube; and one by adding to the cube. And now the secondary figures are generated by subtracting from these two primary figures. Is nothing produced by adding to the secondary figures?

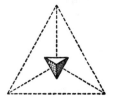

To this second subtraction there also corresponds a second addition to those three primary figures—an angle arising in place of the plane, and a plane in place of the angle. But the figures are the same as those produced by former subtraction. For just as formerly the octahedron was inscribed in the cube, and the icosahedron in the dodecahedron: so now in turn the cube is made to be inscribed in the octahedron, and the dodecahedron in the icosahedron. Accordingly, when all these operations have been performed, the first five figures are found.

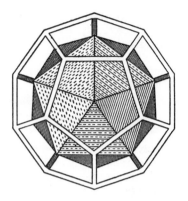

Why do you call them the most simple figures?

Because each of them is bounded by planes of one species alone, *viz.*, triangles or quadrilaterals or pentagons, and by solid angles of one species alone—the three primary figures by the trilinear angle, the octahedron by the quadrilinear angle, and the icosahedron by the quinquelinear angle. The other figures vary [465] either with respect to the angle or with respect to the plane. For there are some which have one genus of planes, as the rhombus in this diagram, but no one genus of solid angles. For the dodecahedral rhombus has six quadrilinear angles and [467] eight trilinear angles; and the thirty-sided rhombus has twelve quinquelinear angles and twenty trilinear angles. There are other figures which mingle diverse planes having uniform solid angles, as the thirteen species of the Archimedean solids.

Why do you call those five figures the most beautiful and the most perfect?

Because they imitate the sphere—which is an image of God—as much as a rectilinear figure possibly can, arranging all their angles in the same sphere. And they can all be inscribed in a sphere. And as the sphere is everywhere similar to itself, so in this case the planes of any one figure are all similar to one another, and can be inscribed in one and the same circle; and the angles are equal.

Is there not some other method by which more figures similar to these can be constructed?

None at all. For a solid angle of any figure is constructed from at least three planes. Accordingly, equilateral triangles can join with triangles, quadrilaterals, and pentagons in order to form a solid angle; quadrilaterals, with triangles; and pentagons, similarly with triangles. But six triangles, or three hexagons, complete a surface, and cannot form a solid angle. But magnitudes larger than these, as three heptagons, or three of any other figure, exceed the sum of four right angles which are laid around the same point in a plane. See the scholium to the last proposition in Book XIII of Euclid and Book II of my *Harmonies*.

Then how are the number of the primary spheres and the intervals of the planetary orbits taken from these figures?

Any solid figure is understood to have two spheres, one circumscribed around it and the other touching the centres of its planes; whence the first view of the solid figure [468] as it were invites some architect to circumscribe and to inscribe spheres. So whatever the ratio of the outer sphere to the inner sphere is, that has been made to be the ratio of the sphere of the upper planet to the nearest lower sphere, between which spheres there is the aforesaid interval.

What are the ratios of the spheres in the single figures?

Let the semidiameter of the circumscribed sphere be 100,000. The ratio of the semidiameter of the inscribed sphere is as follows:

In the cube	57,735	The square is equal to one-third of the square on the radius of the circumscribed sphere.
In the tetrahedron	33,333	One-third of the radius of the circumscribed sphere.
In the dodecahedron	79,465	An irrational part, the square on which is between
In the icosahedron	79,465	two-thirds and three-fifths of the square on the radius of the circumscribed sphere: namely, by the subtraction of the square on the apotome from eleven-fifteenths of the square on the radius.
In the octahedron	57,735	The square is equal to one-third of the square on the radius of the circumscribed sphere.

But the octahedron has at the middle section of itself a square formed by the four lines bounding it in the middle, and if a circle is inscribed in this square, its radius will be 70,711, the square on which is equal to half the square on the radius of the circumscribed circle.

Show now what the place of the sphere of the Earth is among these figures.

The five bodies were distributed into two classes above: into those generated first, and those generated second. The former had a trilinear angle, and the latter a plurilinear. For as Adam was the first-born, and Eve was not his daughter but a part of him—and they are both called the first-made, [469] but Cain and Abel and their sisters are their offspring; so the cube is in the first place, wherefrom have arisen, differently and more simply, the tetrahedron—as it were a rib of the cube—and the dodecahedron, but in such a way that all three remain among the primary figures. The octahedron and the icosahedron, with their triangular planes, are as it were the offspring born of the cube and dodecahedron as fathers and from the tetrahedron as mother; and each of them bears a likeness to its parent.

So the three first figures of the same class had to enclose the circuit of the centre of the Earth and the two figures generated second, as the other class, should be enclosed by the sphere in which the Earth revolves, and so this sphere had to be made a boundary common to both orders, because the Earth, the home of the image of God, was going to be chief among the moving globes. For in this way the nature of being inscribed is kept in the second class and that of circumscribing in the first class. For it is more natural and more fitting that the octahedron should be inscribed in the cube, and the icosahedron in the dodecahedron, than the cube in the octahedron, and the dodecahedron in the icosahedron.

And so in this way the circuit of the centre of the Earth was placed in the middle between the planets; for three planets had to be placed outside, on account of the three primary figures; and two had to be placed inside its circuit—on account of the two figures of the second class—to which the sun is added as a third in the inmost embrace of the centre of the mobile spheres. And so Saturn, Jupiter, and Mars were made the higher planets, and Venus, Mercury, and the sun, the lower. But the moon, which has a private movement around the Earth during the same common circuit of the Earth, is among the secondary planets, as was said above.

What is the order among the three outer figures and what place among the planets does each hold?

The cube is the first of the figures, and therefore it was placed between the two farthest spheres, those of Saturn and Jupiter. In the generation of the figures the tetrahedron follows: therefore, it got [470] the place between Jupiter and Mars. The dodecahedron was the last of the three: therefore, the last place was assigned to it between the orbital regions of Mars and of the Earth.

How do you place the two inner figures?

Although the octahedron has the nature of the cube, of which it is the first parts, and the icosahedron that of the dodecahedron, of which it is the last parts; nevertheless the next place after the dodecahedron did not belong to the octahedron, for two reasons. For, first, the two classes of figures are somehow opposed to one another: therefore it was fitting that the beginning of the placing should be at the opposite termini. But since the first place of the outer figures was judged to be the place which tended the more outward; consequently the first place of the inner figures was judged to be the place which tended the more to the interior towards the centre. Secondly, it was more becoming to the nature of the similar figures; the dodecahedron and the icosahedron, and better suited to their being inscribed within one another, that they should succeed one another very closely, with the circuit or sphere of the Earth coming in between, at which as at a common boundary both classes of figures stop.

Therefore it was caused that the icosahedron should be placed between the orbits of the Earth and of Venus, but the octahedron between the inmost orbits, those of Venus and Mercury. But the sun does not have a sphere in which its centre is carried around; and therefore it is outside the number of the primary moving bodies, but it has in itself the source of the movement outside, the fixed stars have stillness in themselves, and they furnish a place for the moving bodies and contain them.

Is there found between these spheres which you have given to each figure the ratio of the figures?

There is found so much the same ratio that, although there is some very small deficiency, nevertheless no interval between two planets approaches nearer to the ratios of the spheres of another figure [471] than those intervals which have been ascribed with the best of reasons to the two planets.

For you see that as Saturn's sphere had less than twice the diameter of the sphere of Jupiter, and Venus' similarly had less than twice the diameter of Mercury's, namely, five-thirds or eight-fifths; so also in the cube and in the octahedron 100,000 is less than twice 57,775. For if you take three-fifths of 100,000 you will have 60,000; but if you take five-eighths, then 62,500 will be the result. Again, as the sphere of Mars had a very small ratio to the sphere which carries the centre of the Earth, and one nearly equal to the ratio of the sphere of the Earth to that of Venus; so too you will see between the spheres of the dodecahedron and of the icosahedron a very small ratio, namely, that of 100,000 to 79,465. You see, thirdly, that, just as the sphere of Jupiter has a very great ratio to the sphere of Mars, namely triple; so too, in the case of the tetrahedron the diameter of the circumscribed sphere is three times the diameter of the inscribed sphere.

If the intervals approach so nearly to the ratios of the figures, why then does some discrepancy remain?

1. Because the archetype of the movable world is constituted not only of the five regular [solid] figures—by which the chariots of the planets and the number of the courses were determined—but also of the harmonic proportions with which the courses themselves were attuned, as it were, to the idea of celestial music or of a harmonic concord of six voices. Now since this musical ornamentation demanded a difference of movement in any given planet—a difference between the slowest and the fastest movement; and this difference is made by the variation of the interval between the planet and the sun; and since the magnitude or ratio of this variation was required to be different in different planets; hence it was necessary that some very small amount should be taken away from the intervals which are exhibited by the figures as uniform and without variation, and that it should be left to the freedom of the composer to represent the harmonies of movement.

[472] 2. And nevertheless that which the regular solids have of their very own was not neglected in this very small discrepancy. For just as the ratio of the spheres of the tetrahedron is perfect, *i.e.*, rational simply, that of the cube and the octahedron, half-perfect, *i.e.*, rational in square but irrational in length, but that of the dodecahedron and of the icosahedron are wholly imperfect, *i.e.*, absolutely irrational; so also the ratio of the tetrahedral planets imitates the figure almost exactly, *i.e.*, in approximately the extremities of the intervals. But the ratios of the cubic and octahedral planets are less exactly like the figures, because the extreme intervals recede from the figures while the intermediate intervals square with them. But the whole intervals of the dodecahedral and icosahedral planets abandon the ratios of their figures, although they approach no others more nearly. Now see how the longest interval of Mars is almost exactly a third part of the least interval of Jupiter, as in the tetrahedron the [diameter of the] inside sphere is one-third [of the diameter] of the outer sphere: so that in this way if the angles of the tetrahedron are placed in the inmost sphere of Jupiter, the tetrahedral planes somehow touch the farthest sphere of Mars. See again how if the angles of the cube are placed in the sphere of Saturn and the angles of the octahedron in the inmost sphere of Venus, the planes of the figures sink into the regions [*i.e.*, spheres] of Jupiter and Mercury, and do not pass above the total regions but enter in approximately as far as the middle. Finally, see how, if the angles of the dodecahedron are placed in the inmost sphere of Mars and the angles of the icosahedron in the inmost sphere of the Earth, the planes of the figures by no means reach to the inscribed regions of the Earth and of Venus, but that nevertheless no planetary intervals approach more nearly to the least ratios of these figures. Concerning these things, see my *Harmonies* (Book v, Proposition xlix and *passim*), where the causes not only of the exact magnitude of the ratios between two spheres but also of the opposite intervals of any single planet are dug up.

[473] *No other conclusion concerning the interposition of the figures can be drawn from the periodic times, can it?*

All ratios of time in this case are greater than the ratios of their orbits and also greater than the ratios of their figures, as will be unfolded in the second part of this book. Nevertheless the property of the figures can be recognized without difficulty even among the time ratios. For as there are three ratios among the figures—the greatest ratio which is alone, and the middle and the least, which are both found in two cases—the greatest ratio is found in the tetrahedron alone, the middle ratio in the cube and the octahedron, and the least ratio in the dodecahedron and the icosahedron; so the greatest and the solitary ratio of times is found between Jupiter and Mars, approximately that of 6 to 1, nearly 12 years to less than 2 years, an argument for the interposition of the tetrahedron. But between Saturn and Jupiter and between Venus and Mercury the ratio of times is less and in both cases approximately the same, an argument for the interposition of the cognate bodies, in the first case the cube and in the second case the octahedron, which have approximately the same ratios between their spheres. For as the 30 years of Saturn are to the 12 years of Jupiter, so approximately are the 225 days of Venus to the 88 days of Mercury. And finally there is the least ratio of times between Mars and the Earth and between the Earth and Venus, and again it is practically the same in both cases, an argument for the interposition, in the first case of the dodecahedron, and in the second case of the icosahedron, which are bodies cognate and having the same ratio. For as the 687 days of Mars are to the 365¼ days of the Earth, so the 365¼ days are to 194 days, since instead of these Venus has 225 days, a little more, and makes the least time ratio of all. The causes of such a small discrepancy are unfolded in my *Harmonies* (Book v).

You do not have any other evidence, do you, except that from the two classes of figures that the globe [474] *of the Earth has the principal proportionality in location?*

Indeed it is not by chance that the mean interval between the Earth, the middle planet, and the sun is found to be almost exactly a mean proportional between the shortest interval of Mars, the lowest of the higher planets, and the longest interval of Venus, the highest of the lower planets. For, as was said above, the space between Mars and Venus was left for the Earth—left undetermined by the inscribing of figures and open and free, so that in dividing it by the sphere of the Earth, either this ratio or some other, if the other were better, could be expressed. Therefore the mean in the classes of the figures, and the middle wall between the higher and the lower planets had to be a mean geometrically too.

Then what determined this interval which the inscribings did not determine?

Even if there is a certain augmented figure, the aculeate or wedge-shaped dodecahedron, which is taken as determining this interval as accurately as the interval between Jupiter and Mars is determined by the tetrahedron, and the relationship of that imperfect figure with its cognates, the dodecahedron and the icosahedron, is seen not to lack its proper ratio; nevertheless the figures alone do not determine these intervals or any others exactly, but this job was left to the ornament of harmonic movements, which demands a certain amount of freedom in determining these intervals exactly.

4. ON THE RATIOS OF THE PRINCIPAL BODIES OF THE WORLD
TO ONE ANOTHER

Where do you judge that the beginning should be made in investigating the ratios of bodies?

From the Earth (1) as the home of the speculative creature, [475] and (2) of the same image of God the Creator; (3) for we read in the holy *Book of Moses* that in the beginning God created the Heaven and the Earth; (4) moreover, the sphere of the Earth is the middle figure between the planets, and their common boundary and even a geometrical mean proportional between the territories of the higher and the lower planets; (5) finally, the very structure of these ratios cries in a loud voice that God the Creator in fitting the bodies and intervals to the solar body, as to a measure prior to their generation, made His beginning at the Earth.

What do you judge to be the reason for the magnitude of the solar body?

The following things argue that the solar globe is the first of all the bodies of the world in the order of creation, at least in the archetypal order; if not also in the temporal: (1) Moses makes light the work of the first day, and instead of light we can understand the solar body. (2) Above by many votes, the solar body obtained the principate among natural things; why not also in quantity and in time, in which it was created?

Furthermore, the first body, because first, does not acquire any ratio to the

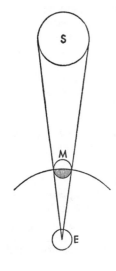

bodies following; but rather the bodies following acquire a ratio to it as first. Wherefore there is no archetypal cause of the magnitude of the sun: nor could there have been a different globe twice as great as it is now, because the rest of the universe and the whole world and man in it would have had to be twice as great as they now are.

Then by what means was the magnitude of the Earth adjusted to the solar globe?

By means of vision of the sun. For the Earth was going to be the home of the speculative creature, and for his sake the universe and world have been made. But now speculation has its origin in the vision of the stars: [476] wherefore too the magnitude of the things to be contemplated had to have its origin in the magnitude of the things to be seen. But the first visible is light or the sun, as it is (1) the work of the first day and (2) the most excellent of all visible things, the principal, the primary, and that which was going to be the cause of the visibility of all the rest. So it follows that the beginning [*principium*] in proportioning the bodies of the world was taken from the vision of the sun from the Earth; just as in the case of the upper planets the regions of the world were separated by the mean proportionality of the orbit of the Earth.

How great does the diameter of the sun appear to be on the Earth?

It is established by the very old observations of Aristarchus and by the most recent of our time, that, if the Earth is at its greatest distance from the sun, and with the point of vision E as centre a circle is described, then exactly one-seven-hundred-twentieth of that circle, *i.e.*, $\frac{1}{2}°$, is occupied and so to speak defined by the diameter of the sun; or, what is the same thing, the angle E, comprehended by the lines touching each rim of S the sun, is one-seven-hundred twentieth of four right angles.

What do you think is the reason for this number?

We must seek the archetypal cause of the first thing among the first causes. But there is no geometrical cause for the division of a circle into 720 parts [477] in a figure lacking that number of sides. For this figure is derived by bisection from a figure having 45 sides; and there is no [geometrical] demonstration of that figure, as is proved in the *Harmonies*, Book I. It follows that this sectioning of the circle is taken from the composition of figures and so from harmonic ratios. And it seems to bring about the necessity that the circle of the zodiac, wherein all the planets had to practise their harmonic movements in reality and the sun in appearance—that this circle, I say, should be divided into parts of a harmonic numbering by the appearance of the first body. But the least number which offers itself in determining all the parts of the monochord and in setting up the twofold scale of the octave [*systema diapason duplex*], *i.e.*, in the minor and the major mode—I say that this number is 720, as was shown in the *Harmonies*, Book III, Chapter 6[1].

Wherefore, since the movements of all the planets, as I show in Book V of the *Harmonies*, had to be adjusted to this twofold scale; it was fitting that the first body, which was the leader of the dance to this music, by its apparent diameter on the Earth, should, in the eyes of the Earth-dweller—*i.e.*, the speculative creature—divide that circle as an index and measure of the apparent harmonic movements, by the division of the monochord, *i.e.*, by the division into 720 parts, which is twice 360, thrice 240, four times 180, five times 124, six times 120, eight times 90, nine times 80, ten times 72, twelve times 60, fifteen times 48, sixteen times 45, eighteen times 40, twenty times 36, twenty-four times 30, by a multiplex form of division into aliquot parts.

Then what follows in respect to the interval between the sun and the Earth from the assumption of this hypothesis? or, what is the magnitude of this rod used by us here-tofore as a measure of the planetary spheres?

If S the diameter of the sun had to occupy $\frac{1}{2}°$ with the point of vision E set up on the Earth; it is necessary that the point of vision, or [478] in place of it E the centre of the terrestrial globe, should be distant from the centre of the sun S

[1]In Kepler's musical system, if the intervals in the octave are represented by assigning whole numbers to the eight tones in such fashion that the ratio between any two numbers will express the analogous musical interval, then the minor scale will be represented by the following sequence, in least numbers:

$$72 : 81 : 90 : 96 : 108 : 120 : 128 : 144$$

Similarly, the major scale, as follows:

$$360 : 405 : 432 : 480 : 540 : 576 : 640 : 720$$

Now if the two scales are combined, "the twofold scale of the octave, *i.e.*, in the minor and the major mode" thus produced will be represented by the following sequence, in least whole numbers:

$$360 : 405 : 432 : 450 : 480 : 540 : 576 : 600 : 640 : 720$$

to the extent of a little more than 229 semidiameters of the round body of the sun S, as we are taught in geometry.

I have the interval; tell me also the magnitude of the terrestrial globe and give reasons.

These things are not yet sufficient to determine the magnitude of the Earth; but it is the work of one axiom added to another. Without doubt, because the Earth was going to be the home of the measuring creature; and the Earth by its body had to become the measure of the bodies of the world, and by its semidiameter the measure of the intervals, as the line is the measure of lines. But since the measuring of bodies is different from the measuring of lines, and since the ratio between the terrestrial body and the solar body is first, as is the ratio between the diameter of the Earth and the interval between the Earth and the sun; nothing is more in agreement with a right and fitting and ordered proportioning than that the equality of both ratios should be postulated—so that as many times as the terrestrial body E is contained in the solar body S, so many times also is the semidiameter of the Earth E contained in SE, the interval between the centres of the sun and the Earth; so that in this way, as the terrestrial body E is to the solar body S, so the semidiameter of the Earth E is to SE, the distance between the centres.

How is the magnitude of the semidiameter of the Earth got from these two axioms?

Let the semidiameter of the sun S be put down as 100,000 parts, so that the interval SE between the centres of the sun and the Earth will be 22,918,166 such parts. The cube of 100,000, *i.e.*, 1,000,000,000,000,000 is to be divided by the interval 22,918,166; and the [square] root of the quotient—which is the continued sine $0°15'0''$ is to be taken, and it will be 6,606. That will be the magnitude [479] of the semidiameter of the Earth. For as 6,606—the semidiameter of the Earth—is contained 3,469⅓ times in 22, 918,166—the interval between the sun and the Earth—so too the cube of 6,606—the semidiameter of the Earth—is contained the same number of times, *i.e.*, 3,469⅓ in the cube of 100,000, the semidiameter of the sun. But it is known from geometry that the ratio of cubes to one another is the same as the ratio of the globes inscribed in the same cubes. So the semidiameter of the sun S will contain the semidiameter of the Earth E a little more than fifteen times; but the body of the sun S will contain the body of the Earth E approximately 3,469 times.

You say that its magnitude is approximately three times what the ancients assigned to the longest distance of the sun from the Earth, for they set a lesser distance, only 1200 times the semidiameter of the Earth: and you say twenty times the [former] ratio of bodies, because they made the sun only 166 times greater than the Earth: do not you respect their astronomical observations?

Not at all. For the ancients made the sun as near as a parallax of 3' should have made it to be. Whence Tycho Brahe reasoned that when Mars is nearer to the Earth than the sun is, Mars should be observed to have a parallax much greater than 3'. But I have observed that the parallax of Mars is not at all perceptible to sense. Therefore the distance of Mars, even when most near, is greater than 1200 semidiameters; and the distance of the sun is greater too.

[480] The diameters of Mars and Venus can be observed through the ancient

instruments and through the Belgian telescope of modern times; and they are
found to be of a very few minutes. Therefore, if the sun is as near as the ancients
say it is; then these planets—each in its own ratio—will become as near as
Tycho Brahe said, out of Copernicus. If Mars is so near; it will be even smaller
in its visible diameter. Therefore Mars will be smaller than the Earth, namely,
the higher planet will be smaller than the lower; so that there will be no proportion [*analogia*] between the magnitude of the bodies and their order—and that
is not in agreement with the adornment of the world.

The greater the distance of the sun is set down to be, the smaller the parallax
of the sun becomes. And the smaller the parallax of the sun, the greater the
parallax of the moon from the sun; if the simple parallax of the moon is taken
from its own principles, and that is of great service in correcting the doctrine of
eclipses. Therefore such a great magnitude for the interval of the sun is confirmed rather than refuted by astronomical observations.

And it accords with the prayer of physics that the solar body, which gives
movement to all the other planets, should be much greater than all the mobile
bodies fused together into one.

*The determination of what body follows most closely upon the determination of the
Earth?*

Of the moon, a secondary planet: (1) because this star has been assigned to
the Earth as its private property, so that the moon might help with the growth
of earthly creatures and be observed by the speculative creature on the Earth,
and that the observation of the stars might begin with it. (2) Because the reasons for the ratio to be established are practically the same.

*Tell the foundations of the ratio between the moon and the
Earth, by reason of the body and by reason of the interval.*

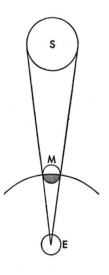

1. Again, the visible diameter of the moon at its greatest distance from the Earth had to occupy $\frac{1}{720}$ of the circle, both on account of the number itself, as before, and
also on account of the eclipses of the sun, a spectacle ordained by the Creator in order that the speculative creature should thereby be taught concerning the rationale of
the course of the stars. And teaching would be done most
rightly if the semidiameters of the sun and moon were to
appear equal at the greatest distance of each: so that in
this way the moon could cover the sun exactly, during
this relationship of both stars, if it occurred. And so M
the moon and S the sun make the same angle at E the
Earth.

2. Moreover, it was fitting that the ratio between the
terrestrial body and the lunar body should be to the ratio
between the distance of the moon and the semidiameter
of the Earth as in the former case the ratio between the
solar body and the terrestrial body was to the ratio between the solar interval and the semidiameter of the Earth: namely, the
proportion holding between the pair of ratios should in both cases be the
same. For the moon, a secondary and terrestrial planet, made in order to

obscure the sun, should have followed the example of the ratios of the sphere of the sun or the Earth.

What follows from this?

Two things follow from postulating the two axioms. And either one of these two things by itself, by a wonderful concord of probabilities, even if it did not follow from the foregoing, [482] could be employed in place of an axiom, since they are very worthy of belief by themselves. The first is as follows: that since the proportion of ratios on the part of the sun is one of equality, *i.e.*, as many times as the body of the Earth E is contained in the greater body of the sun S, so many times is the semidiameter of the Earth E contained in SE the distance or semidiameter of the sphere of the Earth or sun. But that proportion is not more probable than the following: that the body of the Earth E will contain the lunar body M, which is narrower and smaller, as many times as the semidiameter of the Earth E is contained in the distance or semidiameter of the lunar orbit EM, and the former proportion is no more probable than this latter. This very proportion, employed as an axiom, has its own dignity; hence, because the Earth is the home of the measuring creature, therefore the Earth by its own body measures even the smaller lunar body, just as previously it measured the solar body which is greater than it: and by its semidiameter it measures the semidiameter of the lunar sphere. But both proportions are in the ratio of equality, because M the lunar sphere is alone placed around E the Earth, just as the Earth's sphere is placed around the sun; and so the measurement of the lunar sphere and the lunar body, prior to the other planetary bodies, is no less proper to the Earth than previously the measuring of the solar body and the solar sphere was. But in its own proper measuring, it is right for the ratio of equality to hold, as first and principal, if nothing prevents it.

The second thing which follows from the premises by a long circle of demonstration—see it in my *Hipparchus*—is as follows: that by this reasoning the semidiameter of the moon's orbit or the distance EM is a mean proportional between the distance ES, or semidiameter of the Earth's sphere, and the semidiameter of the terrestrial body: so that as the semidiameter of the Earth E is to EM the semidiameter of the moon's sphere, so EM is to ES the semidiameter of the Earth's sphere or the sun's. Here again there is an equality of both ratios which is itself probable, because what the Earth's sphere placed around the sun is to the sun is what the moon's sphere placed around the Earth is to the moon.

[483] *Do observations agree with this interval between the moon and the Earth?*

Down to the last hair: for Brahe found that in its quadratures the distance of the perigeal moon from the Earth was slightly less than 54 semidiameters of the Earth; but the distance of the apogeal moon in the same quadratures is more than 59, slightly less than 60. And from these beginnings it is inferred that the first distance is 54, the second 59.

How is the magnitude of the semidiameter of the moon to be inferred from the axioms and conclusions employed as axioms, which have been posited?

1. If the semidiameter of the moon M is set down as 100,000 parts, so that the interval EM between the centres of the moon and the Earth is 22,918,166

such parts: the cube of 100,000, *i.e.*, 1,000,000,000,000,000, is to be multiplied by 22,918,166; and the fourth root of the product is to be taken: the fourth root will be 389,085, showing the magnitude of the semidiameter of the Earth in terms of the same parts. For just as 389,085 the semidiameter of the Earth is contained in 22,918,166, the distance to the moon, a little less than 59 times, so too the cube of 389,085 will contain the cube of 100,000 slightly less than 59 times, and so the globe of the Earth will contain the globe of the moon slightly less than 59 times. Thus the semidiameter of the terrestrial body E will contain the semidiameter of the lunar body M less than four times.

2. Differently and more simply, from a later conclusion: the square root of 3,469⅓, *viz.*, of the distance to the sun, is taken, and it is slightly less than 59; and EM the distance of the moon is 59 whereof the semidiameter of the Earth is 1. But if 1 the semidiameter of the Earth is divided by 59, and the cube root of the quotient is taken, the result will be the semidiameter of the lunar body in terms of the same dimensions.

Then what is the ratio of the diameters of the sun and moon inferred to be?

The same as the ratio of the solar sphere to the lunar sphere, or that of the [484] lunar [sphere] to the terrestrial body, *viz.*, the ratio holding between slightly less than 59 to 1. And so the solar body contains the lunar body more than 200,000 times.

What are the ratios of the planetary globes to one another?

Nothing is more in concord with nature than that the order of magnitudes should be the same as the order of spheres, so that among the six primary planets, Mercury should have the least body, because it is inmost, and should obtain the most narrow sphere; that next to Mercury should be Venus, which is larger, but still smaller than the Earth, because moving around in a sphere narrower than the Earth's but nevertheless wider than Mercury's; that the larger globe of Mars should be next to the Earth, because its sphere is outside and more spacious, but the lowest of the higher spheres; then the larger globe of Jupiter, in the middle among the higher planets; and finally Saturn's globe, the largest of the moving bodies, because it is the highest.

Now since bodies have three dimensions, along the diameters, across the surfaces, or in the space contained by the surfaces—or the corporeality—the ratio of the surfaces is the ratio of the squares of the diameters, and the ratio of the bodies is the ratio of the cubes of the diameters; it is consonant that one of the three ratios of the globes should have been made equal to the ratio of the intervals. For example, since Saturn is approximately 10 times farther away from the sun than the Earth, either the diameter of Saturn will be 10 times the diameter of the Earth, the surface 100 times the surface of the Earth and the body 1,000 times the body of the Earth: or the surface of Saturn will be 10 times the surface of the Earth, so that the ratio of the bodies will become the ratio of the ³⁄₂ power of the intervals, and thus Saturn will be 30 times larger than the Earth, just as it is 30 times slower; while the ratio of the diameters will be merely the ratio of the square roots of the intervals, *viz.*, Saturn's diameter will be a little more than thrice the Earth's: or the bodies themselves have the ratio of the intervals, so that Saturn will be only ten times greater than the Earth, just

as it is ten times farther away [from the sun], while in the surfaces the ratio of
the $\frac{3}{2}$ power of the intervals will be kept, and in the diameters the ratio of the
cube roots of the intervals. And so the diameter of Saturn's body will be slightly
greater than twice the diameter of the terrestrial body.

Of these three modes, the first is refuted beyond controversy both by the
archetypal reasons and also by the observations of the diameters made with
the help of the Belgian telescope; up to now I have approved of the second mode,
and Remus Quietanus of the third. On my side the better reasons, the archetypal, seemed to stand; on Remus' side the observations stand; but in such a
delicate question I was afraid that the observations were not certain enough
not to be taken exception to.

Nevertheless I yield the place to Remus and his observations. For Jupiter,
opposite the sun and in the perigee of its eccentric circle, was frequently seen
by me to occupy approximately 50″; Remus observes Saturn to occupy 30″;
Mars, opposite the sun and its perigee in the Aquarius, appears greater than
Jupiter, but not much greater. As a matter of fact, if a body equal to the Earth
were seen at as great an interval as we assign to the sun, viz., 3,469 semidiameters
of the earth, it would appear to have a diameter of 2′. But now at the perigeal
distance of Mars, the same body which is equal to the Earth will be perceived to
occupy more than 5′ and thus to be equal to six Jupiters. Therefore, the greater
the diameter of the globe of Mars is than the diameter of the Earth, the larger
will it be in appearance. Therefore, we ought not to make this diameter of the
globe of Mars more than one-sixth greater than the diameter of the Earth, as
takes place in the third mode.

But this argument will make war perhaps not unsuccessfully with the help of
the archetypal reasons: for just as previously we made the ratio of the bodies of
the sun and Earth, and of the Earth and moon, the same as the ratio between
the semidiameter of the Earth and the semidiameters of the spheres, so now
the ratio of the planetary bodies is set down the same as the ratio between the
semidiameters of the spheres. So Saturn in its bodily bulk will be slightly less
than ten times greater than the Earth; Jupiter will be more than five times;
Mars, one and one-half times; but Venus slightly less than three-quarters of the
terrestrial body; and Mercury slightly greater than one-third of the same.

[486] *Just as the body of the Earth, so should not all the planetary bodies be attuned
to the solar body by the same laws as the Earth?*

By no means. For if we were to follow this, the planetary bodies would be
great in an order contrary to the order of the spheres, viz., Mercury would be
greatest; and Saturn smallest with a diameter less than one-third the diameter
of the Earth. That, however, is repugnant both to the aforesaid reasons and to
observations of their diameters. For Saturn opposite the sun, when it is nine
times farther away from the sun than the Earth, occupies approximately 30″;
therefore if it were near to the sun, it would occupy $4\frac{1}{2}′$, since the Earth at an
interval of that size would occupy 2′. And so the diameter of Saturn is more
than twice as great as the diameter of the Earth.

And this is to say, as I said at the very beginning of this section, that it is
very clear from the things themselves that the beginning of setting up the
ratios must be made at the Earth. For observations of the moon and the eclipses
bear witness to the equality of the two ratios, one of which is between the lunar

body and the terrestrial body, and the other between the semidiameter of the Earth and the lunar sphere: we can by no means resist the certitude of the observations. Now it was very probable that the Earth too should be attuned to the sun by the same laws: and when we had set down that, we already had observations agreeing with it distantly: because the observations do not bear out the sun's having a nearness of 1,200 semidiameters of the Earth, but require twice or thrice that distance; and this attunement absolutely demands thrice that distance. Therefore the Earth is certainly the measure both of the solar and lunar bodies and of the solar and lunar spheres. But thus the body of Saturn or any other planetary body can by no means become the measure of both things: once more we bring forward as witnesses of this the certain observations of the diameters. Therefore the Earth alone is the real measure; but the nature of dimensions demands that the beginning of the conformation be taken from the measure.

What should we hold concerning the rarity and density of these six globes?

First, it is not consonant that all the planets should have the same density of matter. For where any multitude of bodies is necessary, there too a variety of conditions is required in order to make distinction, in order that they may be truly many. But the principal condition of bodies as bodies is the internal disposition of the parts. For inequality of bulks somehow happens to the bodies themselves on account of the surfaces bounding the bulks, and the internal part of one body does not differ from the part of another body in this circumscription of bulk. But the principal argument for the dissimilarity of matter is drawn from the consideration of the periodic times: for that consideration will not advance, if we make the globes have the same density, as we shall hear below.

Second, it is consonant that whatever body is nearer to the sun is also denser. For the sun itself is the most dense of all bodies in the world, and its immense and manifold force, which could not exist without a proportionate subject, bears witness to this thing: and the very places which are near the centre wear a certain form [gerunt ideam] of narrowness, such as exists in the condensation of much matter into a narrow place.

Third, nevertheless rarity should not be measured out in proportion to the greatness of the bodies, and density in proportion to the smallness. For example, by the above, both the distance and the amplitude of the globe of Saturn are to the distance and amplitude of the globe of Jupiter as 10 is to 5, approximately. I say that the density of matter of the globe of Saturn must not be put in the same ratio to the density of the globe of Jupiter as the ratio between 5 and 10.

For, if anyone followed this, he would sin against another law of variety by introducing not an unequal amount [copiam] of matter but the same amount throughout all the planets. For when 10 the bulk [488] of Saturn has been multiplied by 5 the density, the product will be 50 the amount of matter, which is as great as the product obtained by multiplying 5 the bulk of Jupiter by 1 its density. But it seems to be preferable and more elegant that the bulks of globes of different density should not be equal to one another, and that the density of globes of unequal bulks should not be the same, and that the amount of matter should not be distributed in equal portions throughout all the globes, which differ in bulk and in the density of their matter. Rather, all these things should vary, so that in whatever order the moving globes succeed one another after the centre, in the same order—order, I say, not ratio—let us measure out not only the spaces between the bodies and the rarity of the bodies but also the amount of matter: thus if Saturn has 50 as its amount of matter, Jupiter should be left with less than 50 but more than 25 the half of that—say 36. For thus the bodies will be as 50 is to 25, the amount of matter as 50 is to 36, the rarity as 50 is to 36, or as 36 is to 25, or the contrary density as 25 is to 36 or as 36 is to 50.

Furthermore, although formerly I upheld the equality of the amount of matter, I have been compelled to assign the ratio of the periodic times to the greatness of the bodies; so that just as Saturn has 30 years and Jupiter 12 years, so too the amplitude of the globe of Saturn to the globe of Jupiter is as 30 is to 12. But the observations of the diameters, made by myself and by Remus, refute this ratio as being too great.

Fourthly, the following things persuade us that the ratio of the amount of matter should be set down as precisely the ratio of the square roots of the bulks or amplitudes—and thus the ratio of the $\frac{3}{2}$th powers of the diameters of the globes and the ratio of the $\frac{3}{4}$th powers of the surfaces. For first, it happens that the ratio of the amount of matter and the ratio of the density are both the ratio of the square roots of the intervals from the sun, and that thus the amount of matter and the density participate equally but inversely in that ratio: hence the amount of matter is greater, hence the density of the same great body is less: and that is the best mean [*mediatio*] of all. For example, Saturn will be twice as high as Jupiter, one and one-half times as heavy, and one and one-half times rarer, or Jupiter is one and one-half times denser. And in comparing the ratios of one planet: Saturn will be twice as high as heavy and twice as ample as rare.

Moreover, this same ratio of the square roots of the intervals is established by the following geometrical propriety: as above, between the intervals of two planets from the sun—for example, let the intervals be 1 and 64 for the sake of easier reckoning—let there be set down two mean proportionals 4 and 16, that is to say, in order to form the two remaining dimensions of the bodies, so that the bodies of the mobile globes would be to one another as 1 is to 64; but the surfaces of the globes as 1 is to 16 or as 4 is to 64, and finally their diameters as 1 is to 4 or as 4 is to 16 or as 16 is to 64. So now between 1 and 64 the intervals of the same two planets from the sun, let there be set down one mean proportional 8; that is to say: in order to form physically the matter within the bodies, for the matter is one thing only; so again let the intervals between the globes be as 1 is to 64, but let the amount of matter and the rarity too be in a lesser ratio, as 1 is to 8 or as 8 is to 64; or inversely the density, as 8 is to 1, or as 64 is to 8. For in this proportion it makes no difference in what manner it is that the corporeality is condensed or rarefied, whether it is merely in length or also in breadth, or finally in all three dimensions. For the ratio introduced prescribes an amount of the thing to be condensed; and different modes of condensation may be found, with the amount always remaining the same.

Therefore if by means of these principles we compute the densities of the planetary bodies and always seek a mean proportional between two planetary intervals from the sun, or, more precisely, between the diameters of two spheres or orbits, and if we finally compare all the numbers to some common round number and reduce them to round numbers: the result will be the numbers which follow in the table; and I have found that the terrestrial matters which I have placed next to them agree fairly closely with them in their ratio—as you can see in my book which I wrote in the year 1616 in the German language concerning weights and measures:

[490]	Saturn	324	The hardest precious stones
	Jupiter	438	The loadstone
	Mars	810	Iron
	Earth	1000	Silver
	Venus	1175	Lead
	Mercury	1605	Quicksilver

So that we may reserve gold—whose density in this proportion is 1,800 or 1,900—for the sun.

Finally what do you set down as the ratio of greatness holding between those three principal regions of the world, between the space wherein is the sun, the space or region of the mobile bodies, and the space of the whole world or the region bounded by the sphere of the fixed stars?

Even if the reasons of Copernicus do not extend to determining by observation the altitude of the sphere of the fixed stars: so that the altitude seems to be like infinity: for in comparison with this distance the total interval between the sun and the Earth, which by the judgement of the ancients embraces 1,200, and by our reasons 3,469, semidiameters of the globe of the Earth, is imperceptible: nevertheless reason, making a stand upon the traces found, discloses a footpath for arriving even at this ratio.

But in the beginning we must glance at the example of the terrestrial and the solar and the lunar spheres, because the ratios of the whole world are derived from the Earth's own proper ratios. And the region described by these three bodies and their courses is as it were a small world. For what the sun is in the Copernican region of the fixed stars—that is, what the Earth is in the sphere or region of the sun—in appearance, at any rate, and for Tycho also is the truth of the thing. And just as the sun is at the centre of the sphere of the fixed stars, unmoving in an unmoving home; so too with respect to the movement of the moon, the Earth is motionless at the centre of the, so to speak, motionless sphere of the sun. For just as the region of the moving bodies is arranged around the sun, so also the sphere of the moon is drawn around the Earth: in the former case the sphere of the fixed stars is the boundary for the planets; here the sun itself is the boundary for the moon, and she returns to this boundary at the end of a month when all her phases are completed.

Therefore it is consonant that, just as by necessary reasons the sphere of the moon was made a mean proportional between the apparent sphere of the sun and the terrestrial body at its centre; so too the region of the moving bodies, or the outmost circle of Saturn, becomes a mean proportional between the outmost sphere of the fixed stars and the solar body at the centre of the world.

Again the same thing is accomplished, even without reference to the small world, by the consideration of the great world itself. For since in one respect the movable bodies strive after the immobility of the encircling body, which supplies the place, while they struggle against the movement, so that they are not moved with such great speed as the mover strives after; in another respect they receive movement from the mover to a certain degree, so that movement from the mover and rest from the body supplying the place are somehow mixed together in the movable bodies. Therefore if it is permissible to state a physical thing in mathematical words, the movable bodies can very aptly be called a mean proportional between the body which is the source of movement, and the immovable body which supplies the place.

But since this thing is true [verum] physically and spatially—for the source is inside, the thing supplying the place is outside, and the movable bodies are in the middle—therefore nothing is more probable [verisimilius] than that even geometrically the semidiameter of the region of the movable bodies should be the mean proportional between the semidiameter of the solar body and the semidiameter of the sphere of the fixed stars, so that just as the solar globe is to the spherical system of all the planets, so this system is to the spherical body of the whole world, which is bounded by the region of the fixed stars. See the diagram on page 854.

How do we know the ratio of the diameter of the solar body to the diameter of the region of the movable bodies?

With the aid of mathematical instruments, from the angle which the solar body occupies in our vision. For since [492] this angle is approximately $\frac{1}{2}°$, it follows that the sun has a distance of 229 of its semidiameters from our vision. But our point of vision is on the Earth; and the diameter of the sphere of the Earth, placed around the sun, is slightly greater than $\frac{1}{10}$ of the diameter of the sphere of Saturn. Therefore the outmost sphere of the movable bodies, i.e., Saturn's, contains approximately ten times as many diameters of the sun [as the terrestrial sphere does], i.e., about 2,000. In the diagram on page 854 this [outmost sphere of Saturn] is the middle circle.

How great does the sphere of the fixed stars turn out to be by this reasoning?

As the diameter of Saturn's sphere, the outmost sphere of the movable bodies, contains the diameter of the solar body about 2,000 times, so too the diameter of the sphere of the fixed stars would contain the diameter of Saturn's sphere about 2,000 times. And so the diameter of the sphere of the fixed stars will contain about 4,000,000 diameters of the solar body, and—according to the ratio between the solar and terrestrial bodies believed by the ancients—more than five times as many diameters of the Earth, i.e., 20,000,000; and by our reasoning, three times that, i.e., 60,000,000.

But is not this amplitude of the sphere of the fixed stars unbelievable, for you make it to be 2,000 times greater than the sphere of Saturn, although among the ancients this sphere stood just above Saturn?

But much more unbelievable is the rapidity—for the ancients—of the sphere of the fixed stars and Saturn. And since it is necessary for one of these two opinions to stand, it is more probable that the sphere of the fixed stars should be 2,000 or 1,000 times wider than the ancients said than that it should be 24,000 times faster than Copernicus said. For in the former case there is no movement present in a subject which is very spacious and as it were infinite; but in the later case an as it were infinite movement is placed in the small sphere of Saturn. In itself such a great amplitude is not repugnant to Brahe's observations, nor is it discordant with reason that bodies which are at rest should be distant from the movable bodies by such an immense interval.

[493] *How do you know that such a great amplitude is not repugnant to Braahe's observations?*

He observed the greatest altitude of the pole star—which at that time was at 7° of the Ram—in the year 1586 at the midnight after the autumn equinox; and it was 58°51'. He observed the greatest altitude [of the same star] on the winter solstice on December 26 at about the sixth hour of the evening, and he found it again at 58°51'. And so there was no difference; even though during the month of September the horizon cut the sphere of the fixed stars lower down—by approximately the whole semidiameter of the sphere in which the Earth is borne—than on December 26, because on the first date the sun was apparent in the Libra; and on the second date, in the Capricornus. The same thing occurred when the least altitude was observed on the midnight after the spring equinox and after the winter solstice at 6 o'clock in the morning; for in both cases the altitude was found to be 52°59½', although during the month of March the horizon cut the sphere of the fixed stars higher up—by approximately the whole semidiameter of the sphere wherein the Earth is borne—than in December. Therefore the diameter of the sphere wherein the Earth is borne is not perceptible through Brahe's instruments.

And so since the diameter [of the terrestrial sphere] does not make a difference of 1' in the sphere of the fixed stars, therefore it is not $\frac{1}{3500}$th part of the semidiameter of the sphere of the fixed stars. Therefore the semidiameter of Saturn's sphere—which is approximately 10 times the semidiameter of the terrestrial sphere—is not equal to $\frac{1}{350}$th part or $\frac{1}{400}$th part of the semidiameter of the sphere of the fixed stars. But it is much less possible to decide whether therefore it is $\frac{1}{2000}$th part, *i.e.*, where the aforesaid altitudes of the polar star differ by $\frac{1}{5}$' or 12", since the diameter of the pole star seems to be equal to at least 1', and since we cannot believe in the diligence of observers to the extent of $\frac{1}{5}$'.

According to Brahe, Saturn has a distance of 12,300 semidiameters of the Earth from the centre of the Earth. Therefore its diurnal circle, when it is on the equator, contains 77,314 semidiameters of the Earth, *i.e.*, 66,420,000 German miles; and their division into 24 hours make the portion of one hour to be 2,767,500; and the sum of 240 miles—for according to Copernicus that was the space traversed by Saturn in one hour—are 1/12,500th part of that.

But according to Ptolemy, by Copernicus' corrections, the ratio of the spheres would be as follows:

The moon is 64⅙ semidiameters distant from the Earth.

50⅚ for the body of the moon and Mercury's

65	lowest point in sphere of Mercury	28½ : 91½
209	highest point	
1	for the body of Mercury and Venus'	

210	lowest point in sphere of Venus	1⅔ : 19⅚
1,407	highest point	
7	for the body of Venus and the sun's	

1,414	lowest point of solar sphere	} 57½ : 62½	Although Copernicus
1,537	highest point		has 1094 : 1190
6	for the solar body		
2	for the body of Mars		

1,545	lowest point in sphere of Mars	} 14½ : 105½
11,241	highest point	
2	for the body of Mars	
5	for the body of Jupiter	

11,248	lowest point of sphere of Jupiter	} 45¾ : 74¼
18,253	highest point	
5	for the body of Jupiter	
5	for the body of Saturn	

18,263	lowest point of sphere of Saturn	} 49⅘ : 70⅕
25,737	highest point	
5	for the body of Saturn	

25,742 This is twice as wide as [495] what Brahe has; and the 240 miles, the hourly movement of Saturn in Copernicus, is smaller than 1/24,000th part of Saturn's hourly movement in Ptolemy.

What do you think are the ratios of density of the solar body, the ether which permeates the whole universe, and the sphere of the fixed stars which encloses all things from the outside?

Since these three bodies are analogous to the centre, the surface of the sphere and the interval, three symbols of the three persons of the Holy Trinity; it is believable that there is only as much matter in one as there is in either one of the two remaining: in such fashion that a third part of the matter of the whole universe should be packed together into the body of the sun, although in comparison with the amplitude of the world the body of the sun is very narrow; that likewise a third part of the matter should be spread out thin throughout the immense expanse of the world; that the sun should in this fashion possess within its own body as much matter as outside of itself the sun is fated to illuminate with the mighty power of light and to penetrate with its rays; and that finally, a third part of the matter should have been rolled out in the form of a spherical surface and thrown around the world on the outside as a wall. And in order that we may shadow forth the proportion to some extent of a known thing which is similar, even if we cannot equal the proportion, let us imagine that the solar body is all gold, the sphere of the fixed stars of water, or glass, or crystal, and the inside space full of air. Whence we are able to understand to a certain extent what divine Moses signified by the Firmament—"raquia," which properly means expansion, *viz.*, the blowing in of the ether—and what by the supercelestial waters. For similarly boys have a game which is an image of creation, when they make bubbles out of soap and water by blowing air into them. The difference is that God holds up the drop—so to speak—of water on the inside [496] at the centre. In the case of the boys, the drop of water, on account of its weight, does not remain at the centre, and is not separated from the surface by the blowing but sticks at the bottom of the bubble.

How great do you set down the thickness to be, or the distance between the inner and the outer surface of the sphere of the fixed stars?

Since we have given it as much matter as is in the total expanse of the world which it embraces; with the exception of the matter which is in the very narrow globe of the sun, but since by no means must the matter of the sphere of the fixed stars be set down as having the same desnsity as the matter in the region of the mobile bodies but a density which is a mean proportional between the density of the ether and the density of matter in the solar body; therefore the sphere of the fixed stars should have an extension which is a mean proportional between the extension [497] of the solar body and the extension of the celestial ether. But, as above, the ratio of the diameter of the sun to the diameter of the ether was as 1 is to 4,000,000. Therefore the ratio of the extensions is the ratio of the cubes, *i.e.*, 1 to 64,000,000,000,000,000,000. But the mean proportional between these numbers is 8,000,000,000. Therefore, that number of spaces equal to the body of the sun will be equal to the space between the concave and convex surfaces of the sphere of the fixed stars. And so the whole world, when its three members have been added together, is represented by the number 64,000,008,000,000,001. And its cube root, $4,000,000\frac{1}{6,000}$, shows that this sphere, having a thickness of $\frac{1}{6,000}$th part of the semidiameter of the solar body and thrown up around the celestial ether, embraces in its body 8,000,000,000 spaces equal to the solar body. Therefore this skin, or tunic of the world, or crystalline supercelestial sphere, is of such great subtlety, on account of the amplitude of its expansion, that if you made it coagulate into one spherical mass, it would have a semidiameter 2,000 times greater than the semidiameter of the solar body, since at present it is not more thick than $\frac{6}{1,000}$ of the semidiameter of the solar body, or a little more than 2,000 German miles.

How great will the sun appear to be if you imagine an eye placed on one of the fixed stars?

The 4,000,000th part of the semidiameter of the [sphere of the] fixed stars subtends about $\frac{1}{20}''$. Therefore the solar body appears to have a diameter of $\frac{1}{600}'$, and it measures the great circle 1,296,000 times; or the apparent diameter of the sun from among the fixed stars is $\frac{1}{18,000}$th part of its apparent diameter viewed from the earth.

How great in turn do the fixed stars appear from the Earth?

Skilled observers deny that any magnitude as it were [498] of a round body can be uncovered by looking through a telescope; or rather, if a more perfect instrument is used, the fixed stars can be represented as mere points, from which shining rays, like hairs, go forth and are spread out.

Does it seem therefore that any one of the fixed stars is such a body as the sun is, and that the sun in turn is seen from the fixed stars to be of so great and of such an appearance as any one of the fixed stars?

I do not think so: for these observations do not prevent the sun from having a body of greater bulk than the fixed stars. Moreover, the view of the sun from such a great interval would be brighter than that of whatever fixed stars. For if, for example, you pierce through a wall with only a pin, so that the sun can shine through the hole, a greater brightness is poured through from the beams than all the fixed stars shining together in a cloudless sky would give. And the eye is not injured by any of the fixed stars; but it cannot bear to look towards the sun even from a distance.

PART II

ON THE MOVEMENT OF THE BODIES OF THE WORLD

1. HOW MANY AND OF WHAT SORT ARE THE MOVEMENTS?

[499] *What was the opinion of Copernicus concerning the movement of bodies? For him, what was in motion and what was at rest?*

There are two species of local movement: for either the whole thing turns, while remaining in its place, but with its parts succeeding one another. This movement can be called δίνησις—lathe-movement, or cone-movement—from the resemblance; or rotation from a rotating pole. Or else the whole thing is borne from place to place circularly. The Greeks call this movement φορά, the Latins *circuitus*, or *circumlatio*, or *ambitus*. But they call both movements generally revolution.

Accordingly Copernicus lays down that the sun is situated at the centre of the world and is motionless as a whole, *viz.*, with respect to its centre and axis. Only a few years ago, however, we grasped by sense that the sun turns with respect to the parts of its body, *i.e.*, around its centre and axis—as reasons had led me to assert for a long time—and with such great speed that one rotation is completed in the space of 25 or 26 days.

Now according as each of the primary bodies is nearer the sun, so it is borne around the sun in a shorter period, under the same common circle of the zodiac, and all in the same direction in which the parts of the solar body precede them [500]—Mercury in the space of three months, Venus in seven and one-half months, the Earth with the lunar heaven in twelve months, Mars in twenty-two and one-half months or less than two years, Jupiter in twelve years, Saturn in thirty years. But for Copernicus the sphere of the fixed stars is utterly immobile.

The Earth meanwhile revolves around its own axis too, and the moon around the Earth—still in the same direction (if you look towards the outer parts of the world) as all the primary bodies.

Now for Copernicus all these movements are direct and continuous, and there are absolutely no stations or retrogradations in the truth of the matter.

By what arguments is it proved that the sphere of the fixed stars does not move?

It was shown in Book I that the sphere of the fixed stars does not rotate around its centre and axis. For we attribute wholly to the Earth whatever appearance of this meets the eyes. Let the other arguments be sought there, in folium 104 *et seqq.* Let us repeat two things alone as proper to this place, one as regards the speed. For if the outmost sphere contains at least 4,000,000 diameters of the sun in its diameter; the circumference will be more than 12,-566,370 solar diameters in length. And if all that revolves in 24 hours, then in one hour 523,600 diameters will revolve; in one minute 8,727; in one second—which is approximately equal to the heart-beat of man—145 diameters of the sun, which is not less than 13,000 German miles; and so during the space of time during which the artery once dilates and again contracts, with a twin pulse-beat, around 7,500,000 (German) miles of the greatest circle would be revolved—and Saturn, in an orbit 2,000 times narrower, would still traverse approximately 4,000 miles.

The second argument destroys completely every movement of the sphere of the fixed stars. For it is not apparent for whose good, since nothing is outside of it, it changes its position and appearances by being moved to what place or from what place, and since it obtains by rest [501] whatever it could acquire by any movement. For the movements of all bodies are understood from its rest; and unless it gives them a place, as it can do perfectly by being at rest, nothing can be moved.

How is the ratio of the periodic times, which you have assigned to the mobile bodies, related to the aforesaid ratio of the spheres wherein those bodies are borne?

The ratio of the times is not equal to the ratio of the spheres, but greater than it, and in the primary planets exactly the ratio of the ⅔th powers. That is to say, if you take the cube roots of the 30 years of Saturn and the 12 years of Jupiter and square them, the true ratio of the spheres of Saturn and Jupiter will exist in these squares. This is the case even if you compare spheres which are not next to one another. For example, Saturn takes 30 years; the Earth takes one year. The cube root of 30 is approximately 3.11. But the cube root of 1 is 1. The squares of these roots are 9.672 and 1. Therefore the sphere of Saturn is to the sphere of the Earth as 9,672 is to 1,000. And a more accurate number will be produced, if you take the times more accurately.

What is gathered from this?

Not all the planets are borne with the same speed, as Aristotle wished, otherwise their times would be as their spheres, and as their diameters; but according as each planet is higher and farther away from the sun, so it traverses less space in one hour by its mean movement: Saturn—according to the magnitude of the solar sphere believed in by the ancients—traverses 240 German miles (in one hour), Jupiter 320 German miles, Mars 600, the centre of the Earth 740, Venus 800, and Mercury 1,200. And if this is to be according to the solar interval proved by me in the above, the number of miles must everywhere be tripled.

2. Concerning the Causes of the Movement of the Planets

[502] *State the opinion of the ancient astronomers as to how the planets move.*

The ancients, Eudoxus and Callippus, and their follower Ptolemy did not advance beyond circles, wherewith they were accustomed to demonstrate the phenomena—not worrying as to how the planets completed these circles: for in Book XIII of the *Almagest*, Chapter 2, Ptolemy writes as follows:

"But let no one judge that these interweavings of circles which we postulate are difficult, on the ground that he sees that for men the manual imitation of these interweavings is quite intricate. For it is not right for our human things to be compared on a basis of equality with the immortal gods, and for us to seek the evidence for very lofty things from examples of very unlike things.

"For is anything more unlike anything than those things which are always in the same state are unlike those things which never stay like themselves, and than those things which can everywhere be impeded by all things are unlike those things which can be impeded not even by themselves? Indeed we must try hard to fit the most simple hypotheses to the celestial movements, in so far as that is possible; but if that is not successful, whatever sort of hypotheses can be used. For if only all things which appear in the heavens are given as a consequence of these hypotheses, then there is no reason for being surprised that interweavings of this sort can occur in the movements of the celestial bodies. For these [interwoven circles] do not have a nature which may impede their movement, but only a nature which has grown fitted to give way and to offer a place for the natural motions of each planet, even if the motions happen to be contrary to one another: so much so that all the circles, speaking absolutely, can interpenetrate all circles with no more difficulty than the movements can be perceived. And these movements occur with ease not only around the single circles, but also around the whole spheres, and around the axes of curved and closed surfaces. For even if the various interweavings of circles, on account of the different movements, and the engrafting of one circle in another are very difficult in the customary representations which are constructed by the human hand, and do not succeed so easily that the movements themselves are not at all impeded: nevertheless we see in the heavens that such a manifold concourse of movements by no means stands in the way of the single movements taking place. Indeed, we should not judge what is simple in celestial bodies by the examples of things which seem to us to be simple, since not even here does the same thing seem to be equally simple in all lands. For it will easily happen that he who wishes to judge celestial things in this way will not recognize as simple any of those movements which take place in the heavens, not even the invariable constancy of the first movement: because it is not only difficult but utterly impossible to find among men this thing (namely, something which stays in the same state perpetually). Therefore we must not form our judgement upon terrestrial things, but upon the natures of the things which are in the heavens and upon the unchanging steadfastness of their movements. So it comes about that in this way all the movements are seen to be simple, and much more simple than those movements which seem to us to be simple. For we are unable to suspect them of any labor or any difficulty in their revolutions."[1] So Ptolemy.

What do you find lacking in this opinion of Ptolemy's?

Even if it is true for many reasons that we should not judge of the ease of celestial movements from the difficulty of the movements of the elements, nevertheless it does not follow that with respect to the celestial movements no terrestrial cases are akin; and Ptolemy seems to draw out this excuse to such lengths that he undermines the whole possibility [*universalem rationem*] of astronomy; and so the excuse satisfies neither the astronomers nor the philosophers, and cannot be tolerated in a Christian discipline.

For as regards astronomy, he brings all hypotheses under suspicion of falsity so long as he argues so strongly for the diversity of celestial and terrestrial things, so that even reason is put down as erring in its judgment [504] of what is geometrically simple. For if that which to our reasoning concerning the heavens seems to be composite, because our reason compounds circles, is simple in the heavens themselves, therefore in the heavens circles are not compounded with one another in order to fashion one movement. Therefore the astronomer is making a false supposition, and, as is extremely astonishing, is eliciting the truth from things which are absolutely false. But that is to destroy the honor of astronomy, which Aristotle upholds in his books of

---

[1] I have, for obvious reasons, translated Kepler's Latin rather than Ptolemy's Greek.
C. G. Wallis.

Metaphysics, believing that "the astronomers should be listened to on the form, lay-out, and movements of the celestial bodies." But in truth Ptolemy reveals himself as regards what he desires: for he says to construct hypotheses which are as simple as possible, if that can be done. And so if anyone constructs simpler hypotheses than he—understanding simplicity geometrically— he on the contrary will not defend his composite hypotheses by this excuse but will say to prefer the hypotheses which seem simpler to us men of the earth, even if we employ terrestrial examples.

As regards philosophy: the philosophers will deny that it is sufficient that the matter of the celestial body should be liquid and permeable by the globes and so should not resist the motions of the globes through it. For they ask what this thing is which leads the globe around, especially if it is established that the matter of the globes resists the movers. They ask by what force the mover moves the body from place to place, as there is no immobile field remaining underneath, and since a round body does not possess the services of feet or wings, by the motion of which animals transport their bodies through the ether, or birds through the air by pressing upon and springing up from the air-current. They ask by what light of the mind, by what means the mover perceives or forms the centres of the circles and the encircling orbits. Finally, neither theology nor the nature of things can bear that Ptolemy, who is steeped in pagan superstition, should make the stars to be visible gods—namely, by inferring immortal life from their eternal motion—and should attribute more to them than belongs to God Himself the Founder—that is to say, that geometrical reasons which are really composite, and the understanding whereof [505] God wished man His image to have in common with Him, should be simple in the stars.

State Aristotle's opinion as to how the planets move in a circle.

Aristotle, believing that the heavens were joined together by solid spheres— though of an equivocal matter—and the later philosophers, whom the Arabs seem to have followed, and after them Peurbach the writer on the schemata— they, I say, at first believed astronomy as regards the number of circles necessary in order to demonstrate the appearances: so Aristotle believed Eudoxus and Callippus concerning the twenty-five spheres. He attributed to the spheres the same number of motor intelligences, who were to revolve in their mind the time of the period and the region of the world into which the motion was to proceed. But since it was probable that all the spheres should look to the same beginning, Aristotle judged that twenty-four other spheres should be placed between these twenty-five spheres, and he called them ανελιττοντες, or counter-turners: namely, in order that each lower sphere should be freed by the interposition of the counter-turner from the carrying off which it was going to suffer from the higher sphere on account of the contiguity of the surfaces. The counter-turners move in an equal time and in a direction opposite to that of the higher sphere, and by that resistance give an appearance of rest, wherein as in an immobile place the lower sphere is stayed and completes its own proper period. And so the mover of each sphere was appointed to give to his own sphere and to all the lower spheres which it embraced a most regular movement within the higher sphere which was placed in contiguity to this sphere. But since that philosopher had decided that movement was eternal, he appointed movers which were also eternal and immaterial, because material things could not have an infinite power. Therefore it followed that the movers were separate and immobile beginnings. But since this eternal duration of the celestial essence seemed to him to be the goodness and perfection of the whole world, as being opposed to destruction, which was something evil; he also gave to these beginnings the highest perfection and the understanding of this perfection, and from understanding the good the will [506] to pursue it, lest [the intelligence] should not do well that which is good; in this way he introduced to us separate minds and finally gods, as the administrators of the everlasting movement of the heavens—just as Ptolemy did. As a matter of fact Scaliger, who professed Christianity, and other followers of Aristotle dispute as to whether this movement of the spheres is voluntary and as to whether the beginning of will in the movers is understanding and desire. And indeed, if the world were eternal as Aristotle contended, at any rate the fixed region in which the planet revolves would bear witness concerning the understanding. For we Christians cannot deny that the highest wisdom has presided over the instituting of the movements whereby the planet is made to run into its own region and is dispatched into its own spaces as if from the barriers; but Aristotle assigned this office to the movers themselves, as being eternal.

Furthermore, motor souls were added, tightly bound to the spheres and informing them, in order that they might assist the intelligences somewhat; or because it seemed necessary for the first mover and the movable to unite in some third thing; or because the power of movement was finite with respect to the space to be traversed and the movement was not of an infinite speed but was described in a time measured out according to space: and that argued that the ratio of the motor power to the movable body and to the spaces was fixed and measured.

And so by this solidity of the spheres and by the constant strength of the motor power absolutely all the movements or celestial appearances were so taken care of, that—given the beginning of movement—then indeed every variation in the movements would arise from the lay-out and plurality of the spheres without any labour or worry on the part of the intelligence; and the spheres moved around poles which were at rest—in approximately the way in which in Book I the terrestrial body was said to rotate around its axis and its poles. And by that movement every sphere—and certain people make them wholly of adamant, so that they by no means yield to any body—carried around its planet, which was bound to the sphere at a fixed place; and one sphere supported another [507] as was said above: and there was no fear that the globes or spheres would fall, bound to one another in this way.

How do you feel about this philosophy?

Again, I do not raise as an objection to it so much the authority of the Christian discipline as the absurdity of the teaching which fashions gods whose functions are among the works of nature and which meanwhile ascribes to them from eternity such things as are necessarily started by one first beginning of all things at the commencement of time. And since this reasoning cannot do without its theology, the whole thing is overthrown by the denial of gods.

Further, solid spheres cannot be granted, as was proved above. But once more, this philosophy rests upon solid spheres, and it is overthrown by undermining them. For Aristotle will readily grant that a body cannot be transported by its soul from place to place, if the sphere lacks the organ which reaches out through the whole circuit to be traversed, and if there is no immobile body upon which the sphere may rest.

Moreover, even if we grant solid spheres, nevertheless there are vast intervals between the spheres. Either these intervals will be filled by useless spheres which contribute nothing to the state of movement; or else, if there are not solid spheres throughout these intervals, then the spheres will not touch one another or carry one another.

Finally this theory abandons itself, in seeing to it that one sphere rests upon another, but forgetting the lowest sphere. For if we are to grant that spheres are supported by spheres and that they are contiguous to one another, then what supports the lowest sphere of the moon or by what columns is it supported upon the Earth, which, as they suppose, is at rest? Since nowhere on the surface of the Earth is any solidity met with: the winds, clouds, and birds freely and easily come and go everywhere. Why doesn't the great weight of the heavens sink down upon us, especially when the denser parts of the spheres [508] approach our zenith? Or if the heavens have no weight, what need do we have of spheres for carrying the planetary globes?

If there are no solid spheres, then there will seem to be all the more need of intelligences in order to regulate the movements of the heavens, although the intelligences are not gods. For they can be angels or some other rational creature, can they not?

There is no need of these intelligences, as will be proved; and it is not possible for the planetary globe to be carried around by an intelligence alone. For in the first place, mind is destitute of the animal power sufficient to cause movement, and it does not possess any motor force in its assent alone, and it cannot be heard or perceived by the irrational globe; and even if mind were perceived, the material globe would have no faculty of obeying or of moving itself. But before this, it has already been said that no animal force is sufficient for transporting the body from place to place, unless there are organs and some body which is at rest and on which the movement can take place. Therefore the question falls back to the above.

But on the contrary the natural powers which are implanted in the planetary bodies can enable the planet to be transported from place to place.

But let it be posited as sufficient for movement that the intelligence should will movement into this or that region: then the discovery of the figure whereon the line of movement is ordered will be irrational. For we are convinced by the astronomical observations which have been taken correctly that the route of a planet is approximately circular and as a matter of fact eccentric—that is, the centre [of the circle] is not at the centre of the world or of some body; and furthermore that during the succession of ages the planet crosses from place to place. Now as many arguments can be drawn up against the discovery of such an orbit as there are parts of it alreayd described.

For firstly, the orbit of the planet is not a perfect circle. But if mind caused the orbit, it would lay out the orbit in a perfect circle, [509] which has beauty and perfection to the mind. On the contrary, the elliptic figure of the route of the planet and the laws of the movements whereby such a figure is caused smell of the nature of the balance or of material necessity rather than of the conception and determination of the mind, as will be shown below.

Finally, in order that we may grant that a different idea from that of a circle shines in the mind of the mover: it is asked by what means the mind can apply this or that [idea] to the regions of the world. Now the circle is described around some one fixed centre, but the ellipse, which is the figure of the planetary orbits, is described around two centres.

Then what seat will you give to mind, so that it may measure out a circle or an elliptic orbit on the liquid plains of the ether? You do not place the mind at the centre, do you? For then you are placing it in the ether, which is not different from all the remaining space of the world, because the orbit of the planet is eccentric to the solar body. But this is exceedingly absurd, since elsewhere the beginning of individuation of souls is assigned to the matter and to the body, to which the soul is added, and this matter differs in place and time and in many other marks from the remaining matter of the world. Surely no other position belongs to the soul and to the mind than that which comes through its body, which the soul informs. And by what force will mind be moved from place to place in a small circle around the centre of the world, so that it may be at the centres of the planetary orbits in the succession of ages, if the mind is without a body and is no more able to be moved than to be given a position in space? By what means will mind view its position or its distance from the centre of the world?

But let it be granted that the mind has a view from its seat at the centre: then how will it cause the planet, which is very distant, to trace its orbit around this centre? If the mind had the planet tied by a rope, perhaps the planet would fly around, being tied to the centre. Perhaps the mind, looking out from the centre, could perceive—especially if it were endowed with bodily eyes—whether the planet were moving in a circle, if the planet were always viewed making an equal angle; but if it should go outside of its circle, in what way would the mind lead it back, if it did not see the orbit by itself? [510] But how does the mind understand the orbit, which is not stamped on the body as its special property? For here there is no question of the intellectual idea of a circle, wherein there is no distinction of great and small, but of the real route of the planet, which has a fixed magnitude in addition to the idea.

But if you place the motor mind outside of the centre of its orbit, its condition will be worse. For either it will be in the body which is at the centre of the world; and thus all the minds will be in the same body, and the above difficulties with respect to keeping the planet in its orbit and with respect to the discovery of the orbit will remain. Or else the mind will be in the globe of the planet: then in both cases it is asked by what means the mind knows where the centre is, around which the orbit of the planet should be organized; and how great the distance of the mind and its globe from that point is. For Avicenna rightly judged that if the mover of the planet is a mind, it has need of knowledge of the centre and of its distance from the centre. For the circle is defined and perfected by the same things, the centre and the equal curvature around it, *viz.*, the distance of the circumference from the centre; and so, however much you exalt the motor mind, nevertheless the circle is nothing else to God except what has already been said. And this same thing should be understood proportionally concerning the figure of the ellipse.

Why do you say that a celestial body, which is unchanging with respect to its matter, cannot be moved by assent alone? For if the celestial bodies are neither heavy nor light, but most suited for circular movement, then do they resist the motor mind?

Even if a celestial globe is not heavy in the way in which a stone on the earth is said to be heavy, and is not light in the way in which among us fire is said to be light: nevertheless by reason of its matter it has a natural ἀδυναμία or powerlessness of crossing from place to place, and it has a natural inertia or rest whereby it rests [511] in every place where it is placed alone. And hence in order that it may be moved out of its position and its rest, it has need of some power which should be stronger than its matter and its naked body, and which should overcome its natural inertia. For such a faculty is above the capacity of nature and is a sprout of form, or a sign of life.

Whence do you prove that the matter of the celestial bodies resists its movers, and is overcome by them, as in a balance the weights are overcome by the motor faculty?

This is proved in the first place from the periodic times of the rotation of the single globes around their axes, as the terrestrial time of one day and the solar time of approximately twenty-five days. For if there were no inertia in the matter of the celestial globe—and this inertia is as it were a weight in the globe—there would be no need of a virtue [*virtute*] in order to move the globe; and if the least virtue for moving the globe were postulated, then there would be no reason why the globe should not revolve in an instant. But the revolutions of the globes take place in a fixed time, which is longer for one planet and shorter for another: hence it is apparent that the inertia of matter is not to the motor virtue in the ratio in which nothing is to something. Therefore the inertia is not nil, and thus there is some resistance of celestial matter.

Secondly, this same thing is proved by the revolution of the globes around the sun—considering them generally. For one mover by one revolution of its own globe moves six globes, as we shall hear below. Wherefore if the globes did not have a natural resistance of a fixed proportion, there would be no reason why they should not follow exactly the whirling movement of their mover, and thus they would revolve with it in one and the same time. Now indeed all the globes go in the same direction as the mover with its whirling movement, nevertheless no globe fully attains the speed of its mover, and one follows another more slowly. Therefore they mingle the inertia of matter with the speed of the mover in a fixed proportion.

The ratio of the periodic times seems to be the work of a mind and not of material necessity.

The most accurately harmonic attunement of the extreme movements—the slowest and the fastest movement in any given planet—is the work of the highest and most adored creator Mind or Wisdom. But if the lengths of the periodic times were the work of a mind, they would have something of beauty, like the rational ratios, duplicate, triplicate, and so on. But the ratios of the periodic times are irrational [*ineffabiles, irrationales vulgo*] and thus partake of infinity, wherein there is no beauty for the mind, as there is no definiteness [*finitio*].

Secondly, these times cannot be the work of a mind—I am not speaking of the Creator but of the nature of the mover; because the unequal delays in different parts of the circle add up to the times of one period. But the unequal delays arise from material necessity, as will be said below, and as if by reason of the balance [*ex ratione staterae*].

Therefore by what force do you suspend your material globes and the Earth in especial, so that each remains within the boundaries of its region, though it is destitute of the bonds of the solid spheres?

Since it is certain that there are no solid spheres, it is necessary that we should take refuge in this inertia of matter, whereby any globe, placed in any place on the world beyond the motor virtues, naturally rests in that place, because matter, as such, has no faculty of transporting its body from place to place.

Then what is it which makes the planets move around the sun, each planet within the boundaries of its own region, if there are not any solid spheres, and if the globes themselves cannot be fastened to anything else and made to stick there, and if without solid [513] spheres they cannot be moved from place to place by any soul?

Even if things are very far removed from us and which are without a real exemplification are difficult to explain and give rise to quite uncertain judgements, as Ptolemy truly warns; nevertheless if we follow probability [*verisimilitudinem*] and take care not to postulate anything which is contrary to us, it will of necessity be clear that no mind is to be introduced which should turn the planets by the dictation of reason and so to speak by a nod, and that no soul is to be put in charge of this revolution, in order that it should impress something into the globes by the balanced contest of the forces, as takes place in the revolution around the axis; but that there is one only solar body, which is situated at the centre of the whole universe, and to which this movement of the primary planets around the body of the sun can be ascribed.

3. ON THE REVOLUTION OF THE SOLAR BODY AROUND ITS AXIS AND ITS EFFECT IN THE MOVEMENT OF THE PLANETS

By what reasons are you led to make the sun the moving cause or the source of movement for the planets?

1. Because it is apparent that in so far as any planet is more distant from the sun than the rest, it moves the more slowly—so that the ratio of the periodic times is the ratio of the 3⁄2th powers of the distances from the sun. Therefore we reason from this that the sun is the source of movement.

2. Below we shall hear the same thing come into use in the case of the single planets—so that the closer any one planet approaches the sun during any time, it is borne with an increase of velocity in exactly the ratio of the square.

3. [514] Nor is the dignity or the fitness of the solar body opposed to this, because it is very beautiful and of a perfect roundness and is very great and is the source of light and heat, whence all life flows out into the vegetables: to such an extent that heat and light can be judged to be as it were certain instruments fitted to the sun for causing movement in the planets.

4. But in especial, all the estimates of probability are fulfilled by the sun's rotation in its own space around its immobile axis, in the same direction in which all the planets proceed: and in a shorter period than Mercury, the nearest to the sun and fastest of all the planets.

For as regards the fact that it is disclosed by the telescope in our time and can be seen every day that the solar body is covered with spots, which cross the disk of the sun or its lower hemisphere within 12 or 13 or 14 days, slowly at the beginning and at the end, but rapidly in the middle, which argues that they are stuck to the surface of the sun and turn with it; I proved in my *Commentaries on Mars*, Chapter 34, by reasons drawn from the very movement of the planets, long before it was established by the sun-spots, that this movement necessarily had to take place.

What do you think should be held concerning the solar body and the force whereby it turns around its axis?

It was said in the first Book that this body—or any other which revolves around its own axis—was not merely moved in a gyre by the Creator's omnipotence at the commencement of things, but it also seems to continue this movement by the reinforcement of a motor soul. For even if, by means of some other rationale there unfolded, the movement could be continued, nevertheless the dailiness and yearliness of this movement, in which the total life of the world consists, is more rightly obtained by the reinforcement of a soul.

Do you have any other arguments besides movement which make it likely that a soul is present in the solar body?

1. A strong argument is drawn from the matter of the solar body and its [515] illumination, which seems to be a quality in the solar body, sprung from the very mighty informing by a soul—namely, a soul whose matter, it is consonant, as was said above, is the most dense matter among the bodies of the world: therefore it is right to believe that very great forces are present in that soul, which dominates and sets on fire the enduring matter.

2. Moreover, I think a soul must be postulated rather than an inanimate form, because it is apparent from the rising of sun-spots and their dispersion and from the unequal illumination of different parts in different times that there is not one continuous and perpetually uniform energy [*energiam*] in all the parts of the solar body, but that it admits movement and variation and interchanges, and that such things take place in the solar globe as take place in the terrestrial globe, *mutatis mutandis*, so that from its inmost bowels hither and yon things which look like clouds are breathed out—which are perhaps black soot—and when their matter has been consumed, the light of the parts which before were covered by those spots becomes more bright. And since these interchanges are perennial, they smell of the guardianship of a soul rather than a simple form.

3. Moreover, light in itself [*per se*] is something akin to the soul: no less than this same thing was proven of heat in Book I. For on the earth nothing is set on fire [*inflammatur*], that is to say, is made luminous, which was not engendered by some soul in a body: as a trunk from the soul of the shoot, alcohol from the soul growing in the vine, sparks from iron and stones, which are things cooked up in the bowels of the Earth by the soul of the Earth. But that light is something akin to our flames is clear from the fact that light concentrated by concave or convex lenses sets on fire in such fashion that there are flames and coals. And so it is consonant that the solar body, wherein the light is present as in its source, is endowed with a soul which is the originator, the preserver, and the continuator.

4. And the function of the sun in the world seems to persuade us of nothing else except that just as it has to illuminate all things, so it is possessed [516] of light in its body; and as it has to make all things warm, it is possessed of heat; as it has to make all things live, of a bodily life; and as it has to move all things, it itself is the beginning of the movement; and so it has a soul in itself.

You don't, do you, further add mind or intelligence to the sould of the sun, in order for it to regulate this movement of the sun around the axis?

There is absolutely no need of mind for the functions of movement. For the region in which the sun revolves exists from the first commencement of things. But the constancy of the revolution and of the periodic time, as was explained above, depends upon the ratio of the constant power of the mover to the obstinancy [*contumaciam*] of the matter. But the directing of the solar axis perpetually towards the same region is stillness rather than the work of mind, since from the first commencement of things none of this movement has been impressed upon the axis. But also the mean circle between the extremes of the axis, the poles, necessarily follows the direction of the axis; and it, regulated by the same perpetually fixed points, abides as the axis abides. Finally, the laying hold of the planetary bodies, which the rotation of the sun makes to revolve, is a bodily virtue, not animal, not mental.

And let these things be said with respect to movement. However, with respect to the inferences concerning intelligence, to which the consideration of the celestial harmonies leads, see the last chapter in Book v of my *Harmonies.*

Then does the sun by the rotation of its body make the planets revolve? And how can this be, since the sun is without hands with which it may lay hold of the planet, which is such a great distance away, and by rotating may make the planet revolve with itself?

Instead of hands there is the virtue of its body, which is emitted in straight lines throughout the whole amplitude of the world, and which—[517] because it is a form of the body—rotates along with the solar body like a very rapid vortex; moving through the total amplitude of the circuit—whatever magnitude it reaches to—with equal speed; and the sun revolves in the narrowest space at the centre.

Could you make the thing clearer by some example?

Indeed there comes to our assistance the attraction between the loadstone and the iron pointer, which has been magnetized by the loadstone and which gets magnetic force by rubbing. Turn the loadstone in the neighbourhood of the pointer; the pointer will turn at the same time. Although the laying hold is of a different kind, nevertheless you see that not even here is there any bodily contact.

The example is certain, but obscure; explain what that virtue is and of what genus of things.

Just as there are two bodies, the mover and the moved, so there are also two powers, by which the movement is administered; one is passive and verges more towards matter, namely the likeness of the planetary body to the solar body, with respect to the bodily form; and there is one part of the planetary body which is friendly to the sun, and the opposite part is unfriendly. The other power [*potentia*] is active and smells more of form—that is to say, the solar body has the force (*vim*) to attract the planet with respect to its friendly part and to repulse it with respect to its unfriendly part, and finally to keep it, if it were placed thus, so that it does not direct either its "friendly" or its "unfriendly" part against the sun.

How can it be that the whole planetary body is like or akin to the solar body, but one part of the planet is friendly to the sun and the other part is unfriendly to the sun?

Doubtless, too, since the loadstone attracts the loadstone, the bodies are akin; but the attraction takes place with respect to one part alone; and the repulsion, [518] with respect to the opposite part. Therefore friendliness and the unfriendless are named from the effect of rushing together or of flying apart, not from the unlikeness of the bodies.

Whence comes this diversity of the opposite parts of the same body?

In loadstones the diversity comes from the situation of the parts in the whole. For if you break the loadstone *AB* at *CD*, then wheresoever the pieces are transposed to, parts *A* and *CD* of the two pieces are mutually repellent.[1] In the whole loadstone these parts formerly looked towards the same region of the world. But if the pieces are put next to one another, so that the former relative situation of the parts occurs, as *CAD*, *BCD*, then the pieces attract[2] one another.

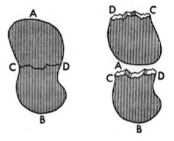

In the heavens the thing is arranged somewhat differently. For the sun possesses this active and energetic faculty of attracting or repulsing or retaining the planet, not as a loadstone does, in one region, but in all the parts of its body. And so it is believable that the centre of the solar body corresponds to one extremity or region of a loadstone, but the whole surface to the other region of the loadstone. Therefore in the bodies of the planets, that part or extremity which at the first commencement of things and at the first placing of the planet [519] looked towards the sun is akin to the centre of the sun and is attracted by the sun; but the part which stretched out away from the sun towards the fixed stars came to possess the nature of the surface of the sun; and if it is turned towards the sun, the sun repulses the planet.

In order that I may better understand the effect of the whirling-movement of the sun, say what you judge would have been the case if the sun did not have a whirling-movement.

Just as a loadstone does not stop attracting a loadstone which has its friendly part turned towards it, until it brings the second loadstone into bodily contact and unites it completely to itself; and if the unfriendly part is turned towards it, either it turns the second loadstone around and in the same way attracts the loadstone which has been turned around; or else, if it cannot turn the loadstone

[1]*Reading* repellunt *for* attrahunt.
[2]*Reading* attrahunt *for* repellunt.

around, it repels it, and in this case the [first] loadstone does not leave any place to the second within the sphere of its virtue, if only it is not hindered. So we must consider in the case of the sun, that if it did not turn around its axis, none of the primary planets would revolve around the sun; but a part of the planets would voyage towards the sun perpetually, until they were united to it by contact; and the part which turn their behind towards the sun would be repulsed towards the fixed stars; but the planets which show their side to the sun would stick to their place and be utterly immobile, while the attractive virtue of the sun struggles with the repulsive.

Then what takes place now by the sun's rotating around its axis?

Indubitably by the turning of the solar body the virtue too is turned, just as by the turning of a loadstone the attractive force of one part is transferred to different regions of the world. And since by means of that virtue of its body the sun has laid hold of the planet, either attracting it or repelling it, or hesitating between the two, it makes the planet also revolve with it and together with the planet perhaps all the surrounding ether. Indeed, it retains them by attraction and repulsion; and by retention it makes them revolve.

If this were the case, would not all the planets make their periodic returns at the same time as the sun?

Yes, if only this were the case. But it has been said before this, that besides this motor force of the sun there is also a natural inertia in the planets themselves with respect to movement: hence by reason of their matter they are inclined to remain in their own place. So the motor power of the sun [*potentia vectoria*] and the powerlessness or material inertia of the planet are at war with one another. Each has its share of victory: the motor power moves the planet from its seat; the material inertia removes its own, *i.e.*, the planetary, body somewhat from those bonds by which it was laid hold of by the sun, so that it is laid hold of first by one part and then by another part of this circle of virtue

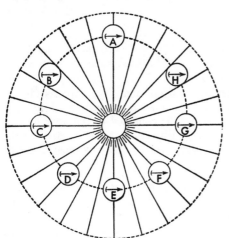

[*circularis virtutis*] and as it were circumference of the sun—that is, by the part which comes next after the part from which the planet has just loosed itself.

In the diagram, the form [*species*] of the rotated solar body is understood by the outer circle designated by dots; [521] and such a circle is understood to be drawn through any position whatsoever of the planet *A, B, C, D, E, F, G,* or *H.* Let the sun, and with the sun its form, turn from right to left: first, let the planet *A* be laid hold of by that part of the form of the sun which is designated by the ray *A*; let the ray *A* be moved through a fixed interval of time as far as the place of the ray D; and let it draw the planet, which however resists and tries to loose itself. Thus, as in the same interval of time the planet is propelled from A to B, so the first ray leaves the planet behind it by the space BD. But in turn, the ray H has succeeded it and lays hold of the planet in B. For as far as A was moved forward to D, so far has H proceeded to B.

But if all things are effected by natural powers which are set at work and at war to move the inertia of matter, how can the planets preserve their periodic times, so that the periodic times are always exactly equal?

More easily than by the assistance of mind. For as the ratio of the total motor virtue to the matter of the globe to be moved is invariable, it follows that the periodic times are perpetually equal.

But why does one planet free itself from this grasp more completely than another, so that Saturn moves only 240 miles in one hour while Mercury moves 1200 miles according to Copernicus?

1. Because the virtue flowing from the solar body has the same degrees of weakness in different intervals which the intervals themselves have, or the amplitude of the spheres described with these intervals: that is the mightiest cause.

2. Moreover there is something of a cause in the inertia of the planetary globes or in the greater or less resistance, whereby the ratio only half corresponds; but more of this somewhat later on.

The body of the planet is always the same but it is repelled by the sun and attracted to it, as you wish; therefore it passes through different degrees of motor virtue: therefore the ratio of the power to the planetary body does not remain constant.

Not absolutely, if we consider the parts of one revolution; and so the same planet becomes faster in one part of its revolution, as above in E, than in another part A, as will be said below. But in spite of this the general total motor virtue [*collecta universa virtus vectoria*] throughout all those degrees through which the planet passes during one revolution always and at every restitution is of the same magnitude.

How is it possible that the virtue flowing from the body of the sun should be weaker in the greater interval at A than near the sun at E? What weakens the virtue or makes it feeble?

Because that virtue is corporeal and partakes of quantity: wherefore it can be dispersed and thinned out. Therefore since as much power is diffused throughout the very wide orbital circle of Saturn as is collected in the very narrow orbital circle of Mercury: therefore it is very thin throughout the parts of the orbital circle of Saturn, and hence it is very feeble; but it is most dense at Mercury and hence is very strong.

If it were a question of the body of the sun, I might grant to it this natural power of moving: but you draw out this material power from the body and place it without a subject in the very spacious ether. Doesn't this seem absurd?

[523] That it should not seem absurd is clear from the example of the loadstone, to which this same objection can be made. But in neither case is this force without a proportional subject. For in this way at the very source the sub-

ject of the natural faculty is the body of the sun, or the threads stretching out
from the centre to its circumference; thus even in this very emanation, I think
a rational distinction should be made between the immaterial form [*speciem*] of
the solar body, which flows as far as the planets and beyond, and its force or
energy which actually lays hold of the planet and moves it—so that the form is
the subject of the force, though it is not a body but an immaterial form of a
body.

Could you give an example of this thing?

There is a true example in the light and heat of the sun. There is no doubt but
that just as the whole sun is luminous, so it is all on fire, and that on account of
the density of its matter it should indeed be compared to a glowing mass of
gold, or to anything else which may be denser. Now from that light [*ex luce illa*]
of the sun there emanates and comes down to us a form which is not corporeal,
not material, which we call the illumination [*lumen*] or rays of the sun and which
however is subject to dimensions and accidents. For it flows on straight lines
and may be condensed or rarefied, and many indeed be cut by a mirror and by
glass, namely, by reflection and refraction, as we are taught in Optics. Moreover, this form of the sun's light [*lucis*] bears its heat with it; and in proportion
to the greatness or smallness of the strength whereby it falls upon bodies
which can be illuminated, it warms them to a greater or to a lesser extent.

Therefore just as that form (*species*) or illumination (*lumen*)—which form we
know with certainty to flow down from the light (*luce*) of the sun—is the subject of the heat-giving faculty, which has similarly been extended from the sun,
through a form; so too the solar body's immaterial form, come down as far as
the planets, has as its companion the form of that energetic virtue (*speciem
illius virtutis energeticae*) in the solar body; and this form strives to unite like
things to itself and to repel unlike.

There is a more clear example in this same light: when it passes through
coloured glass or coloured weavings or has been communicated to coloured surfaces, [524] it itself is coloured. Hence, it cannot be denied that, although the
light is an immaterial form of the light which flowed into the coloured body, it
becomes the subject of that colour and as it were even the outward-going
vehicle of it.

*What if this very light and not some other form from the body of the sun were the
subject of the prehensive faculty whereby the sun lays hold of the planetary bodies?*

Not absolutely; for it seems rather that we should take it that an immaterial
form flows off from its body, in which form the prehensive force and the light
inhere, but in the light the colour and heat—each of them drawn from its own
source.

*State the reasons for this distinction between the immaterial forms from one and the
same solar globe.*

1. The matter of the solar body must be something distinct from the light
in it. For the movement of rays of light in a straight line takes place in an instant; but the turning of the solar body takes place in time. But if we were to
postulate that the bare form of the light is the subject and vehicle of the prehensive virtue, then one and alone the light of the sun could lay claim to being

the whole essence of its body. For the same thing which is found in the form from the thing is found in the thing as in its origin [*orginaliter*].

2. Dimensionality applies to emanated light doubtless not wholly by reason of the inward essence of the light, but by reason of something different from the light itself, namely, because the light is in a body of some certain magnitude and because the forms emanate together as much from body as from light.

3. The form of light emanates from the surface of the luminous body, or if really from the depths of a pellucid body, still as if from the surface. And so the light is considered as a surface, [525] and it has the same properties which other surfaces have with respect to movement and impact; but the body which is beneath the illuminated surface suffers nothing, because nothing has come down from the inward corporeality of the source of light: now the force laying hold of a body must necessarily come down from a body, so that there may be a moving cause which is proportional to its movable object. And so it is subject to bodily dimensions and moves bodies: not only with respect to the surface but also winding its way into their very matter.

4. Hence, too, no matter on the surface of the object resists the light in such a way that the surface is not illuminated instantaneously; but what resists the light, *i.e.*, something opaque, resists it perpetually and is never overcome, as long as it remains opaque. But the virtue which lays hold does not overcome every whit: for the resistance of matter in the planetary body stands up against it and restricts it: hence the planet does not follow exactly the forward movement of the prehensive force, but is left behind and abandoned by it and in that mutual struggle there is place for time.

5. There is the same reason for the further difference that light is bounded and stopped by the surfaces of opaque bodies, so that it goes on no farther into other bodies lying in the same straight line. But this force which moves the planet by laying hold of it is not stopped by its surface, but goes into the body which it lays hold of, and moreover goes on through the body into the body of a farther planet, if it so happens that two planets are on a straight line with the sun: consequently the movement is not disturbed at all by the interposition of bodies. But if movement arose from the illumination of light, this would be absurd; for as often as the higher planet was eclipsed by the lower, so often and so long would its movement cease until the lower planet by its speed should remove itself from the line [with the sun].

6. Finally, that the planetary movement is not necessarily from the naked light of the sun is shown by the examples of other things where movement similar to the celestial movements takes place without light, as can be seen in the case of the loadstone; and this will be shown below by the example of the moon, which is moved by the earth, a body which is luminous to the least extent. And even if at that time the illumination of the earth and the moon has a share [526] and even if it co-operates in many ways in moving the moon, nevertheless it does not do that *per se*, but merely strengthens the motor form from the earth (*specimen motricem telluris*), as will be said in the proper place.

What is the likeness between the form of light and the form of this prehensive virtue?

There is a very close likeness in the genesis and conditions of both forms: the descent of each from the luminous body takes place instantaneously; each remains of average greatness and smallness without loss, is not taxed; nothing perishes in the journey from its source, nothing is scattered between the source and the illuminable or movable thing.

Therefore each is an immaterial outflow, not like the outflow of odours, which are conjoined to a decrease of the substance; not like the outflow of heat from a raging furnace, or anything similar, by which the spaces in between are filled. For this form is not anywhere except in the opposite and withstanding body; the form of the light on its opaque surface, but the form from the motor virtue in the total corporeality: but in the intermediate space between the sun and the surface, the form is not but has been. But if they were to meet the concave spherical surface of an opaque body, both solar forms would be scattered in that concavity together with all that abundance with which they have emanated from the body of the sun: in this way as much of the form would be in a wide and farther-away sphere of this sort as is in the narrow and nearer sphere. And since the ratio of convex spheres is the ratio of the squares of their diameters: therefore the form will be made weaker in unequal spheres in the ratio of the square of its distance. And again because circles have the same simple ratio as their diameters: therefore in longitude the form is weaker in the same ratio of its distance from its source.

Where are the arguments for this comparison gotten from?

These properties of light have been demonstrated in optics. [527] The same things are proved by analogy concerning the motor power of the sun, keeping the difference between the works of illumination and movement and between the objects of each. And these same things are found to be consonant with astronomical experiments.

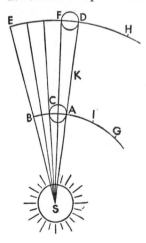

For since one and the same planet, as will be said below, in parts of the eccentric circle which are really equal but are at unequal distances from the sun, makes unequal delays and does that in this very ratio of the distances therefore it follows that the motor virtue is attenuated in longitude in the same ratio wherein the light is attenuated in longitude, namely in the ratio of the amplitude of the circles which have those distances or semidiameters. In this diagram, let the sun be *S*, the same planet *CA* as nearer and *FD* as farther away. And let *DH* and *AI* be equal parts of the eccentric circle—that is to say, at opposite positions on the eccentric circle. Let *DH* be the farther away, and *AI* the nearer. Accordingly as *SD* is to *SA*, so the delay of the planet in *DH* is to the delay of the planet in *AI*. From this it also follows that as *SD* is to *SA*, so inversely is the density of *CA*, the light which is at a lower distance, to the density of *FD*, the light which is farther away.

[528] *But if the light is attenuated in the ratio of the squares of the intervals, i.e., in the ratio of the surfaces; why therefore does not the motor virtue too become weaker in the ratio of the squares rather than in the simple?*

Because the motor virtue has as subject a form from the solar body, not according as it is merely body but according as it is set in motion in a revolution around its immobile axis and poles.

Therefore even if the form from the solar body is attenuated in longitude and in latitude no less than the light is; nevertheless this attenuation contributes towards the weakening of the motor virtue only by reason of the longitude: for the local movement which the sun gives to the planets takes place only in lon—gitude, wherein even the parts of the solar body are mobile, not also in latitude towards the poles of the body with respect to which the sun is immobile.

But nevertheless the movable bodies have latitude no less than longitude. Wherefore they are borne by this virtue so that they each have their latitude as well as longitude. Therefore why is not this motor virtue weakened in latitude, and that in the ratio of the squares of the intervals?

Indeed the planetary bodies have not only these two dimensions but also the third dimension of thickness or altitude; and they occupy this virtue clearly in three ways: and exactly for that reason the prehensive, vehicular, and motor virtue of one planet is not one circle lacking latitude, but is con-stituted of an infinite number as it were of circles parallel in latitude and in altitude. But it does not accordingly follow that the attenuation of this virtue should be in the ratio of the squares or cubes of the intervals or semidiameters. For just as elsewhere in geometry equimultiples have the same ratio: so also here in physics, as one least physical line—as a part of the planetary body—is to the thinness of one circle of virtue, in the simple ratio of the intervals; so is an infinity of least physical lines—as all the parts of the planetary body laid out in latitude as well as in altitude—to the same number of circles of the motor virtue, which all together and singly have force to move merely in longitude; but neither singly nor taken all together do they have any force to move in latitude or in altitude. Therefore, just as the single lines or solitary threads of two planetary bodies would be moved by the single circles of motor virtue in the simple ratio of the intervals; so, taken together, all the threads of the planetary globe are moved by all the circles of motor virtue, taken together, in the same simple proportion: for the latitude and altitude of the motor virtue is not of its essence but of the accidents of the movable thing.

Nevertheless are not these statements concerning the form from the solar body and from the solar virtue, which makes the planets and the Earth in particular revolve, more difficult to believe than the former statements of the philospohers concerning intelligences, motor souls, and solid spheres?

Their being more difficult to believe is no disadvantage, provided they are easy to comprehend; and the objection made against the spheres and the intelligences cannot be made against them—or any other objection wherein the charge of impossibility is made.

For in the first place, wherever they exceed belief there is none the less a true example in the loadstone. Then, if anyone should doubt whether the faculties of a loadstone, *i.e.*, terrestrial faculties, are present in the heavens, or whether the Earth, a heavy body, could be transported from place to place by an immaterial form from the sun; let him regard the moon, which is so much akin to the Earth: he sees it revolve without any solid sphere underlying it. [530] But that bodily forms which pass one another back and forth are strong enough to cause movement is shown by this same moon, which moves the seas on the Earth by the form given out. So we are not lacking in examples. And the mode by which we perceive in mind what sort of form it is does not disturb us: alone the unbelievable strength [*fortitudo*] of this form keeps us in doubt. And indeed we can rightly answer here with Ptolemy that it is never proper to appraise the virtues of divine works according to our weakness, or their greatness according to our smallness.

Now the appraisal of mode and figures belongs to mind; but in this appraisal there should be no judgment of greatness or smallness, *i.e.*, of indefinite quantities.

4. On the Causes of the Ratio of the Periodic Times

In the beginning of this consideration of movement you said that the periodic times of the planets are found to be quite exactly in the ratio of the 3⁄2th powers of their orbits or circles. I ask what the cause of this thing is.

Four causes come together in establishing the length of the periodic time. The first is the length of the route; the second is the weight or the amount of matter [*copia*] of matter to be transported; the third is the strength of the motor virtue; the fourth is the bulk [*moles*] or space in which the matter to be transported is unrolled. For as is the case in a mill, where the wheel is turned by the force of the stream, so that, the wider and longer the wings, planks, or oars which you fasten to the wheel, the greater the force of the stream pouring through the width and depth which you will divert into the machine; so too

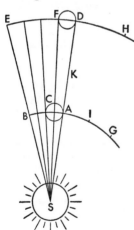

that is the case in this celestial vortex of the solar form moving rapidly in a gyro—and this form causes the movement. [531] Consequently the more space the body—*A* or *D* in this case—occupies, the more widely and deeply it occupies the motor virtue, as in this case *BCA* understood according to its width; and the more swiftly, other things being equal, is it borne forward; and the more quickly does it complete its periodic journey.

But the circular journeys of the planets are in the simple ratio of the intervals. For as *SA* is to *SD*, so too is the whole circle *BA* to the whole circle *ED*. But the weight, or the amount of matter in the different planets, is in the ratio of the 1⁄2th powers of the intervals, as was proved above, so that always the higher planet has more matter and is moved the more slowly and piles up the more time in its period, since even before now by reason of its journey it would have wanted more time. For with *SK* taken as a mean proportional between *SA* and *SD* the intervals of the two planets; as *SK* is to the greater distance *SD*, so the amount of matter in the planet *A* is to the amount in the planet *D*. But the third and fourth causes balance one another in the comparison of the different planets. But the simple ratio of the intervals plus the ratio of the 1⁄2th powers constitute the ratio of the 3⁄2th powers of the same. Therefore the periodic times are in the ratio of the ⅔th powers of the intervals. Consequently if SD, SK, SA, and SL are continued [mean] proportionals, then SL will be to SD [532] as the time period of planet A is to the time period of planet D.

Prove, in comparing two planets, that the weakening of the motor virtue is exactly balanced by the amplitude wherewith the movable planetary bodies occupy the virtue.

The bulks or expanses of the bodies are in the simple and direct ratio of the intervals, as was demonstrated above. That is, as SA is to SD, so is the bulk of the planetary body at A to the bulk of the other planet at D. But too the motor virtue is dense and strong, in the simple ratio of the intervals but inversely; for as the same interval SA is to SD, so the strength of the form CA is to the strength of the form FD. Therefore the virtue is in turn occupied in the same ratio in which it is weakened; for example, Saturn is borne by a power ten times feebler than the virtue by which the Earth is; but conversely, it occupies with its body ten times more of the virtue of its region than the Earth with its body occupies of the virtue of its region. And let the total virtue which Saturn occupies by its bulk be divided into ten parts which are equal in expanse (*spatio*) to the total virtue which the Earth occupies. Any one of these parts or expanses of virtue has only one tenth of the strength which that one part which is occupied by the Earth has, wherefore those ten parts added together into one are equal in power [*potestate*] to that one part by which the Earth is borne. And so if in the amplitude of the more rarefied globe of Saturn there were not more matter than in the narrowness of the denser terrestrial body, the globe of Saturn in one year would be borne along as great a distance of its orbit as is the length of the whole orbit of the Earth; and thus in ten years it would complete its proper orbit. But in fact it has approximately thrice as much matter and weight as the Earth does; wherefore it requires thrice as long a time, namely thirty years.

[533] *What need was there to teach of this balancing? Would it not have been enough for establishing a demonstration to set down that there is absolutely no cause for such an irregular movement as this either in the different grades of the motor virtue or in the different amplitudes of the planetary globes?*

Now for the demonstration that the ratio of the different periods of the planets is the ratio of the ⅔th power of the intervals, it would have made no difference whether this or that were set down. But if we had progressed to the different delays of one and the same planet at different intervals, we could not have established from the same genus of things the reason why the delays in arcs which are exactly equal should follow the ratio of the intervals.

Then what is the reason why the farther distant from the sun any equal arc of the eccentric circle is, the longer delays does the planet make in that arc, and in the ratio of the intervals?

The reason is indeed the weakening of the motor virtue: just as light is SD the longer interval from the sun is diffused more thinly along the length FD than is the diffusion of the same in the shorter interval SA. And so what of the virtue was at that time occupied by the body of the planet, as FD, is more weak than what of the denser virtue is occupied by the same CA which is nearer.

For here the three remaining causes are missing. For the arc or route is assumed in both cases to be of the same length, as *DH*, *AI*: the density of the body remains the same, and the magnitude of the figure likewise; because *FD* and *CA* are in this case one and the same planet. The strength of the virtue alone is left. But more on this in the following.

[534] *In this case we seem to meet a greater difficulty than above. For when the planet is nearer to the sun, it occupies not only longer arcs of the circles of the motor virtue but also denser arcs: wherefore should it not extend its delays in the ratio of the squares rather than in the simple ratio of the intervals?*

Now the same thing is said as above, and the same answer is made. For even if Saturn at that time did not come down into the sphere of the Earth: nevertheless we were comparing the expanse of power occupied by Saturn not merely with that which the Earth would have occupied in the sphere of Saturn but also with that which the Earth would occupy in its own sphere. Therefore, as before, the fact that the circles are denser (*confertiores*) is to be assigned to the form from the body; and this form is something distinct from the inhering motor virtue, which extends in longitude alone and gets no advantage from the condensation of its subject in latitude—unless a thin line without width (*latitudine carens*) has no natural force in length (*in longum*): where the width (*latitudo*) of such a line is judged not by the density but by the expanse, namely on account of the width of the bodies to be moved, as I taught above.

5. ON THE ANNUAL MOVEMENT OF THE EARTH

Accordingly this philosophy of Copernicus makes the Earth one of the planets and sets it revolving among the stars: I ask what, besides what has been said, is required for the easier perception of the teaching and the arguments.

Since the annual movement of the Earth becomes necessary, because it has been postulated that the centre of the sun is at rest at the centre of the world, and since this movement is caused [535] by the revolution of the sun in that space and clearly removes the truth of the stopping and retrogradation of the planets and explains it as a mere deception of sight: we must distinguish carefully the following questions: (1) Whether the sun sticks to the centre of the world. (2) Whether all the five spheres of the planets and the middle sphere of the Earth are drawn up around the sun, so that the sun is in the embrace of all. (3) Whether the sun occupies the very centre of the whole planetary system, or whether it stands outside of that. (4) Whether this centre of the system and the sun in it revolve in an annual movement, or whether the Earth has an annual movement through the sections opposite to those in which the sun is thought to be moving at any time.

You have proved above that (1) *the sun is at the centre of the sphere of the fixed stars. Now prove also that* (2) *it is within the embrace of the planetary spheres.*

That the sun is at the centre of the planetary revolutions is proved first from an accident of this movement, namely, the appearance of stoppings and retrogradings, which are deceptions of sight, or even because the planets seem to be faster when progressing than they really are.

For to begin with the lower planets, now for a long time during the many ages

which followed Ptolemy—let us say nothing at present of ancient Aristarchus—
it was perceived by the authors Martianus Capella, Campanus, and others, that
it is not possible for the sun, Venus, and Mercury to have the same period of
time, namely, a year, unless they have the same sphere and unless the sun is at
the centre of the two spheres of Venus and Mercury and these planets revolve
around the sun: for that reason, when these planets seem to retrograde, they are
not really retrograding but are advancing in the same direction in the sphere of
the fixed stars but are going around the sun. And that is more consonant with
the nature of celestial things.

[536] A few years ago Galileo confirmed this argument by a very clear demonstration: by means of a telescope he disclosed
the illumination of Venus. When Venus is pro-
gressing and is in the neighbourhood of the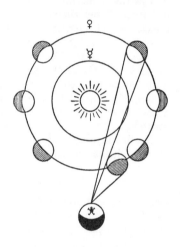
sun, it has a round figure; when retrograding,
a horn-shaped figure. For from this it is proven
with the utmost certainty that its illumination
comes from the sun, that when Venus appears
round and progresses straight ahead, it is
above the sun; but that when it is horn-shaped
and retrograding, it is below the sun, and that
it thus revolves around the sun. Let the demonstration of this thing by reason of light be
joined to the demonstrations of the illumina-
tions of the moon. In the case of Mercury,
Marius brings forward similar things by the aid
of the telescope: the feebleness of its light was
recognized as the planet came down to the
Earth: and that is a sign that the form
(*speciem*) of the illumination is changed and
the light has become weaker in the horn, so that it moves the eye less when near
than when far away. And that would be absurd without this weakening in the
horn; because elsewhere things which are nearer appear greater than if they had
drawn away farther. [537] Now as regards the three upper planets, Aristarchus,
Copernicus, and Tycho Brahe demonstrate that if we set them in order around
the sun and put the sun as, so to speak, the common centre of the five planets, so
that the movement of the sun, whether true or apparent, affects all the spheres
of the five planets, we are freed—as before in the case of Venus and Mercury,
from two eccentric circles, so now in the upper planets—(1) from three epicycles;
(2) from the blind unbelievable harmony of their real movement with the movement of the sun; (3) and just as above in the case of Venus and Mercury, they
have no real stations and retrogradations with respect to the sun, around which
they revolve; (4) thus also very many complications in the latitudinal movement are removed from the doctrine of the schemata; (5) and finally the reasons
are disclosed for the difference which makes the five planets become stationary
and retrograding, but never the sun and the moon; and (6) why Saturn, the
highest of the upper planets, has the least arc of retrogradation; Jupiter, the
middle one, the middle arc; and Mars, the nearest one, the greatest arc. All these
things will be explained below in Book vi. But the ancient astronomers were
totally ignorant of the reasons for these appearances.

3. But even the secondary planets bear some witness to this thing. For Marius found that in the world of Jupiter the restitutions of the jovial satellites around Jupiter is never regular with respect to the lines which we cast out from the centre of the Earth to Jupiter; but that they are regular, if they are compared with the lines drawn through Jupiter from the centre of the sun. And that argues strongly that the orbit of Jupiter is arranged around the sun and that the distance of the sun from the centre of the orbit of Jupiter is sure and somehow fixed; but that throughout the year the Earth varies its distances from this centre.

How many sects of astronomers are there with respect to this theory, from which the second argument is drawn?

Three: the first, commonly known by the name of the ancients, [538] nevertheless has Ptolemy as its coryphaeus; the second and third are ascribed to the moderns. Though the second, named after Copernicus, is the oldest, Tycho Brahe is the founder of the third.

Accordingly Ptolemy treats the single wandering stars only separately, and he ascribes the apparent causes of all the movements, retrogradations, and stations, to their separate spheres. None the less in the case of each planet he sets down one unchanging sphere which completes its period with respect to the movement of the sun: but Ptolemy does not explain from what causes that comes about—unless the Latin writers, hypnotized by their complete ignorance of the rays, attribute some obscure force to the constant rays of the sun.

The remaining two founders compare the planets with one another; and the
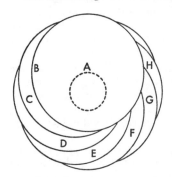
things which are found to be common to their movements are deduced from the same common cause. But this common cause, which makes the planets seem stationary and retrograding in some fixed configuration of the planet and the sun, is still attributed by Brahe to the real movement of all the planetary spheres; but it is completely removed by Copernicus from the spheres of the planets. For Brahe teaches that all the five spheres of the primary planets are bound together at some common point, which is not far distant from the centre of each sphere— as if here all the spheres were described in the common circular table *B*—and that this, so to speak, common node or knot really revolves with the sun during the year, and that very near to this node the sun—in the small circle made with dots—makes all the spheres revolve along with itself; and so to speak dislocates them from their own regions in the world—in the manner wherein bolters, grasping a sieve by one part of its rim, turn it with their hands and shake it; hence, for example, the position of the whole planetary system during the month of June is along circle *B*, during August along *C*, during October along *D*, during December along *E*, during February along *F*, during March along *G*, whence once more along *B*: meanwhile the planet, which has not been disturbed at all by this dislocation of its sphere, [539] completes its own circle on the sphere around

its, as it were, fixed centre. But as regards the time of one year, Copernicus leaves the centres of the spheres absolutely fixed and also the centre of the sun fixed in the neighbourhood of the aforesaid centres. But he ascribes to the Earth, and thus to our eyesight, an annual movement around the sun: hence, since our eyesight thinks that itself is at rest, the sun seems to move with an annual movement and all the five planets seem now to stop, now to go in the opposite direction, now to advance forward very fast.

So by what arguments do you prove (3) *that that common node or common node or centre of all the primary systems is not* [merely] *very near to the sun but is in the body of the sun at its centre?*

In the teaching of astronomy the following arguments for this thing are drawn:

1. From the movement of altitude and longitude of the planets. [540] Observations duly taken bear witness that the longest line in the schema of each and any primary planet, which exactly bisects the orbital circle into two semicircles equal in the magnitude and speed of their parts, passes through the centre of the sun. Therefore all the five lines of altitude always meet at the centre of the sun. See the diagram on page 22.

2. From the movement of latitude of the planets. For we learn from the same class of things, that is, from observations, that the orbit of each and any primary planet is cut by the ecliptic at the positions opposite the centre of the sun, not opposite any other neighbouring point.

3. But if the centre of the sun and the centre of the region of the movable bodies were different, then the very slow movement in the small circle would have to be ascribed either to the centre of the sun or to the centre of the region of the movable bodies, on account of the progression of the apogee of the sun, as will be taught in Books VI and VII. And thus the one of these two which moves could not either be or remain at the centre of the world. But it is probable that both of them are at the centre of the world and are at rest there: the sun does so on account of the preceding and the following arguments. But the node of the movable bodies does so by reason of being the source of movement, and we have already said that this movement starts from this common centre of the movable bodies. But stillness belongs to the source; and on account of stillness, a place at the centre of the movable bodies and at the centre of the whole world.

4. The seat to be assigned to this same source of movement is not in any mathematical point, very near to the most noble body, but rather in that most noble body, for three reasons: first, in order for us to avoid the absurdity that the source of movement—which is necessarily set down as being at that common node of all the spheres, as will be proved below—should be very near to the heart of the world, but nevertheless should not be at the very heart of the world, namely the sun; secondly, because the motor force cannot reside in a mathematical point but requires a body, namely the heart of the world, the sun; thirdly, because the motor force absolutely demands for itself the centre of the world, where the sun itself is: just as stillness belongs to the surface of the world, so movement belongs to the inside.

5. [541] But in especial, the following thing must be taken out of the reach of Brahe's judgement and be demonstrated; that the centre of the region of

the movable bodies does not differ from the centre of the sun. For if Brahe admits this, he will be forced to assign some movement to the sun; and he will be forced to admit that besides the centre of the sun, there is another, different centre for the movable bodies; and by this movement it comes about that the sun now precedes this centre, now follows it, now stands above, now stands below; and nevertheless both always have the same period of time.

(6) As a matter of fact, something absurd and surprising would happen to Brahe. For the sun would be moved by an eccentric movement, having its apsis today in the Cancer. But the centre of the movable bodies would have the apsis of its eccentric movement in the opposite sign of the Capricornus. But what would be the reason for this thing?

(7) These two last arguments furnish an argument against Copernicus also inasmuch as he places that common node of the planets very near to the sun, but not in the sun itself. For the movements of all the remaining primary planets agree in the fact that the points around which their movements appear regular are different in position from the common centre of the region of the movable bodies: the Earth alone would keep this point as the measure of its movement, if the sun were not in the centre of the region of the movable bodies. But what would be the cause of this difference?

(8) Finally the reason why Copernicus and Brahe make these two centres different is not sufficient nor astronomical enough. For they were led to that by the fact that they wished in the forms of their hypotheses to express their every-way equipollence to the Ptolemaic form. But it was not necessary that they should step exactly in Ptolemy's foot-prints. For indeed Ptolemy did not build up all the parts of his hypothesis from observations, but he based many things upon the preconceived and false opinion that it is necessary to presuppose that the movements of the planets are regular throughout the whole circle—and that is demonstrated by observations to be false. Let anyone who wishes fully to understand these astronomical arguments which are placed here under one aspect—let him go to my *Commentaries on the Movements of the Planet Mars.*

[542] *Finally by what arguments do you prove* (4) *that the centre of the sun, which is at the midpoint of the planetary spheres and bears their whole system—does not revolve in some annual movement, as Brahe wishes, but in accordance with Copernicus sticks immobile in one place, while the centre of the Earth revolves in an annual movement?*

Even though the other necessarily follows from the demonstration of the one, nevertheless certain arguments pertain more closely to the sun and certain to the Earth; and certain others equally to both.

First on this side was the same argument whereby we just now claimed for the sun the midpoint of the spheres: namely, that the superfluous multitude of spheres and movements has been removed. For as it is much more probable that there should be some one system of spheres of the sun and that it should be common to the centre of the sun and to that node of the five spheres, according to Tycho Brahe, than that we should believe according to Ptolemy that in any one of the five planets, over and above the spheres which have to do with their proper movements, there is present one whole system of spheres exactly like the sixth system of the sun; so also it is now much more probable that the centre of one Earth should revolve in an annual movement and the sun be at

rest, according to Copernicus, than that, according to Brahe, this node of the
five systems together with the spheres and planets themselves and the sun as a
sixth should have the same annual movement besides the other movements
which are proper to each. For even though Brahe removed from the true systems of the planets those five superfluous schemata of Ptolemy, which are like
those of the sun, and reduced them to that common node of the systems, hid
them, and melted them down into one; nevertheless he left in the world the very
thing which was effected by those schemata: that any planet, over and above
that movement which must really be granted to it, should be moved by the
movement of the sun and should mix both into that one movement. And since
there are no solid spheres, [543] from this mixing there are caused in the expanse
of the world very involved spirals. See the diagram of this involution in my *Commentaries on Mars*, folium 3.

Copernicus on the contrary by means of this one simple movement of the
centre of the Earth stripped the five planets completely of this extrinsic movement of the sun, and made the centres of the six primary planets—that is, the
Earth and the remaining five—each describe singly a simple and always similar
orbit, or line very close to a circle, in the expanse of the world.

The second argument is from the movement in latitude. If epicycles revolve
around an Earth at rest, either according to Ptolemy or according to Brahe; it
will be necessary for those epicycles, especially those of the lower planets, in
different ways to seek the sides as well as the head and feet, that is, to have a
twofold libration. But with the Earth in motion, all the orbital circles have a
constant inclination to the ecliptic. See Book vi, Part iii where the latitudes of
the lower planets supply us with a very clear argument for the movement of the
Earth.

Thirdly, just as above, in the doctrine on the sphere, the diurnal revolution
of the Earth being granted, the immense sphere of the fixed stars was freed
from a diurnal movement of incalculable speed; so now, an annual movement
being granted to this same Earth after the model of the other planets, we have
ended that very slow movement of the fixed stars, which is called by Copernicus
the precession of the equinoxes. See Book vii as regards these things. For it is
much more believable to attribute them to the axis of the Earth, a very small
body, than to such a great bulk.

Fourthly, the consideration of the ratios of the spheres wars on this side.
For it is by no means probable that the centre of a great sphere should revolve
in a small sphere. For the proper spheres of the three upper planets are much
greater than the sphere of the sun—Saturn's approximately ten times greater;
Jupiter's five times; Mars' one and one half times.
Therefore these five spheres are not carried around or
dislocated from their position; but their centres remain approximately fixed, and, as a consequence,
instead of this movement common to them and to the
sun, the Earth revolves.

[544] The fifth argument, which is related to the
preceding one, is the same as that whereby Brahe
tried to disprove the solidity of the spheres. For if
Brahe's reasoning holds, as the orbit of Mars is one
and one half times the orbit of the sun, so the body of

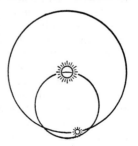

Mars at fixed times returns to that point in the world's expanse where the sun was at other times. And it is quite unbelievable that the regions which the primary planets pass through should be so jumbled together; since in Copernicus they are not only distinct, but are kept separate by very large intervals of emptiness.

I make the sixth argument similar to the fourth: from the magnitude of the movable bodies. For it is more believable that the body around which the smaller bodies revolve should be great. For just as Saturn, Jupiter, Mars, Venus, and Mercury are all smaller bodies than the solar body around which they revolve; so the moon is smaller than the Earth around which the moon revolves; so the four satellites of Jupiter are smaller than the body of Jupiter itself, around which they revolve. But if the sun moves, the sun which is the greatest, and the three higher planets which are all greater than the Earth, will revolve around the Earth which is smaller. Therefore it is more believable [545] that the Earth, a small body, should revolve around the great body of the sun.

The seventh reason is drawn from the reasons for the intervals, which were unfolded above in the first part of this book. These reasons are disturbed and maimed, unless we grant to the Earth too its own sphere, which Copernicus gives to it between the spheres of Mars and of Venus. For even if the interval between Saturn and Jupiter could be deduced from the cube, that of Jupiter and Mars from the tetrahedron, and that of Venus and Mercury from the octahedron, even in Brahe's ordering: yet there would still remain between Mars and Venus a single interval. But there remain two figures in the number of figures of the world. And the interval between Mars and Venus, which is in a greater ratio than double would not square with one of these figures, the dodecahedron or the icosahedron; nor could it be deduced from two figures, not even by the interposition of some sphere between them.

Eighthly, the same things are to be said concerning the harmony of the celestial movements, which are made up of the same numbers and proportions as our musical scale. And if you consider the excellence of the work or the pleasantness of contemplation, or finally the unavoidable force of the persuasion, this harmony can truly be called the soul and life of all astronomy. But this harmony is at last complete only if the Earth in its own place and rank among the planets strikes its own string and as it were sings its own note through a variation of a semitone: otherwise there would be no manifesting of its semitone, and that again is the soul of the song. As a matter of fact, if the semitone of the Earth is gone, there is destroyed from among the celestial movements the manifesting of the genera of song, i.e., the major and the minor modes, the most pleasant, most subtle, and most wonderful thing in this whole discussion. But concerning this in the *Harmonies*.[1]

Ninthly, if we consider the force of Brahe's ordering and if we image for ourselves some matter of the five dislocatable spheres, which matter together with the region of the movable bodies is dislocated by the annual movement; then in this matter, in this, I say, celestial sphere which extends throughout all the planetary regions, the Earth even when at rest [546] will describe such an orbit around the sun as Copernicus assigned to it between the spheres of Mars and Venus—the sun and the centre of the region of the movable bodies being at rest.

[1]See pp. 169 ff.

And so by an absurd and unfitting reason a joke leads us from a distance to the same beauty—namely, that the Earth should progress along one circle or sphere, while being at rest. It is more believable that the sixth orbit of the Earth is described by the real movement of the very Earth, just as the remaining five orbits are described by the same number of movements.

The tenth argument, taken from the periodic time, is as follows: the apparent movement of the sun has 365 days, which is the mean measure between Venus' period of 225 days and Mars' period of 687 days. Therefore does not the nature of things shout out loud that the circuit in which those 365 days are taken up has the mean position between the circuits of Mars and of Venus around the sun: and thus this is not the circuit of the sun around the Earth—for none of the primary planets has its orbit arranged around the Earth, as Brahe admits—but the circuit of the Earth around the resting sun, just as the other primary planets, namely Mars and Venus, complete their own periods by running around the sun?

The eleventh argument is taken from the motor causes—by the supposition of Brahe's opinion, though it is not admitted by all. For because there are no solid spheres, therefore the motor faculties can be placed nowhere except in the movable bodies. But thus the condition of the motor souls will become mighty hard, and harder that of the intelligences, while the motor souls are ordered to transport the body, in which they are, from place to place by a twofold movement, without the resistance of anything, but the intelligences are ordered to look to very many things—so that they may carry the planet in its own rank by two in all respects distinct and mixed-together movements. For at least at one and the same moment they are forced to look to the beginnings, centre, periods, and figures of either movement. But if the sun is at rest and the earth is moved, each planet has only one movement, and that movement can be effected by bodily magnetic powers. For there is barely need of the animal faculty for the one revolution of the solar body, but there is absolutely no need anywhere for the supervision of mind. See my *Commentaries on Mars passim*.

The twelfth is from the source of movement. For it has just been demonstrated in many ways, and it will be confirmed below, that every movement of the primary planets, and in part even the secondary planets, arises from the sun. But it is right to believe that the first cause of movement is immobile. Therefore the sun sticks immobile to its own place—and as a consequence the Earth moves in an annual movement, in place of the sun.

The thirteenth is from motor instruments. For if we let the sun and the Earth revolve around their own axes, then the forms from these bodies (*horum corporum species*) become the subjects of the motor powers by which the six planets are moved by the sun, and the moon by the Earth. But if the sun revolves in an annual movement while the Earth is at rest, then no form from a body, which form should introduce movement, is at hand for moving the sun. And if the Earth does not revolve around its axis once a day, it does not have anything whereby it may move the moon. But this argument urges the daily movement more strongly [than the annual].

The fourteenth argument is from the movement in longitude. If the sun moves, carrying around with it the system of all the spheres, something novel occurs in the case of the sun. For some body will move itself, or else it surely will be moved by some special outside mover, since the other primary bodies are moved by one common sun, and thus by something other than themselves. But if the Earth moves in a circle and is also moved by the sun, like the other primary bodies, nothing novel happens. And so it is probable that the Earth is moved, for a probable cause of its movement appears; and it is probable that the sun remains fixed.

The fifteenth argument is from the movement in altitude. It has already been said in part, and it will be demonstrated below more fully, that all the planets have a movement of libration in a straight line, which proceeds towards the sun and by means of this libration obey the laws of their speed and slowness in any position on the eccentric circle. And thus it is certain that the sun is the cause of this variation in all five planets. But it has been demonstrated in the *Commentaries on Mars* that the same thing takes place on the Earth, if it moves, [548] namely that the Earth too has a libration along its diameter in the direction of the sun. But if it is put down that the sun is moving, then on the contrary the Earth becomes the cause of the sun's slowness and speed and thus of its revolution too. But indeed let the bodies of the sun and the Eearth themselves be viewed, and let the judgement be made as to whether it is more probable that the sun, which is the source of movement of the five planets and is many times greater than the Earth, should be moved by the Earth, or whether on the contrary the Earth, one among the primary planets, should be moved by the common source of movement of the remaining planets. See the *Commentaries on Mars*.

The sixteenth probability is as follows. Now in Book I it was maintained by many arguments and by the refutations of their contraries that the Earth has a diural rotation around its axis, and among those arguments not the weakest were as follows: If the Earth is put down as having a diurnal movement, the final and instrumental cause of the obliquity of the ecliptic can be taken from this same Earth; and, with the Earth at rest, neither of these [causes] can be explained, nor can they be sought from the sphere of the fixed stars, wherein the zodiac is, without a glance at this petty little body which is called Earth. Therefore the transportation of the centre of the Earth can no longer be absurd. But the probability is sufficient, if the remaining arguments demand the thing itself. For this is not to be offered as a necessary argument: because even though the sun rotates around its own axis, nevertheless it is immovable in place, as a whole.

The seventeenth reason: If the Earth revolves in an annual movement, not only do we find a more probable cause for the precession of the equinoxes than if we assign this variation to the sun, the first body; but also by this same means we give a reason for the irregular progression of the planetary nodes, and by the inclination of the axis of the diurnal movement of the Earth we explain the causes of the change in the obliquity of the ecliptic, just as in the case of some irregularity in the precession of the equinoxes—which irregularity we disapproved by the third argument. But one must necessarily be profoundly ignorant of the causes of so many of these phenomena, if the Earth does not revolve in an annual movement.

[549] Let the eighteenth argument come from the end of movement, by which it is proved that movement belongs to the Earth as the home of the speculative creature. For it was not fitting that man, who was going to be the dweller in this world and its contemplator, should reside in one place of it as in a closed cubicle: in that way he would never have arrived at the measurement and con-

templation of the so distant stars, unless he had been furnished with more than human gifts; or rather since he was furnished with the eyes which he now has and with the faculties of his mind, it was his office to move around in this very spacious edifice by means of the transportation of the Earth his home and to get to know the different stations, according as they are measurers—*i.e.*, to take a promenade— so that he could all the more correctly view and measure the single parts of his house. Now you understand that—in order that the first part of this Book IV might be fitted properly together—its writer needed to have the Earth a ship and its annual voyage around the sun. But if the Earth moves, the sun is necessarily at rest.

6. ON THE DIURNAL REVOLUTION OF THE TERRESTRIAL BODY AROUND ITS AXIS AND ITS EFFECT IN MOVING THE MOON, AND ON THE MUTUAL PROPORTIONS OF THE YEAR, MONTH, AND DAY

Because, in addition to the annual revolution around the sun, the diurnal rotation too is assigned to the Earth, which is one of the primary planets: I ask, you do not believe, do you, that all the primary planets turn in this way around their axes?

That is very probable, firstly in the case of Venus, as being seen to exhibit one spot after another—taking as a sign its sparkling, which is of a different form from the sparkling of the fixed stars; [550] again in the case of Jupiter, as bearing four satellites; and Saturn, which bears two just as the Earth bears one, the aforesaid moon: concerning which satellites below.

By what principles is this rotation of bodies around their axes brought to pass?

In Book I, which concerns the Earth, and in this Book IV, which concerns the sun, it has been said that these bodies are turned by an inborn animal principle or something similar. But that this principle is not alone in rotating the Earth but is assisted by the sun is gathered from two pieces of evidence: first, because the number of daily revolutions of the Earth in a year, which is 365¼, exceeds the proximate archetype, which is 360. For it is fitting that, unless the internal motor force of the Earth were nourished by the perpetual presence of the sun, the Earth would have moved along somewhat more slowly around its axis; and thus in the same space of a year it would have made fewer revolutions, namely only 360. With this postulated, it follows that the 5¼ remaining and as it were supernumerary revolutions are added to those 360 on account of some assistance from the sun. The other piece of evidence names this circumstance: that that part of the difference of time which the preceding Books I and III, folia 108 and 286, spoke of, and which Tycho Brahe is seen to have brought to light by clear experiments with eclipses, and which I reduced to physical form, is relevant here. For because this additosubtractive difference of time puts the summer revolution of the Earth as slightly slower than the winter: that indeed could not come from an inborn principle in the Earth, as such principles are wont to be perpetually uniform; but it must come from the intervals between the sun and the Earth, which in our hemisphere are longer in the summer than in the winter.

Perhaps every force causing this whirling movement is in the sun alone, none in any principle of movement inborn separately in the earth?

This is repugnant to both of the aforesaid reasons. For (1) if [551] the number 365¼ were not composed of the two effects of two distinct causes, there would be no reason why it is not one of the archetypal numbers, that is, one of the round numbers rather than one of the disjointed and ignoble fractions. (2) If the true physical additosubtractive difference of time is postulated: then, if the sun caused everything, the whole diurnal revolutions of the Earth would be proportional to the intervals between the sun and the Earth. But the magnitude of this additosubtractive difference of time demands that not the whole revolutions but merely some small parts of the revolutions should be proportional to those variable intervals.

You reckon the internal virtue of the Earth at 360 revolutions in one year; what reason for this number do you have to show from the archetype?

Because the sun had to cover the 720th part of its circle or apparent course at its farthest distance from the Earth; I believe that such great strength has been united to this virtue of the whirling movement, that the sun in any revolution of the Earth could appear to have progressed through two such small parts of its circle—according to the number of the two parts of the revolution, of which one part is called day, and the other is called night—to the view from any one place on the surface of the Earth. Consequently if two spaces of the ecliptic are stamped by the positions of the sun on two successive noons, there would be intercepted an empty or unstamped space equal to either of them; and as day is to night, so would the space filled by the sun be to the empty space— the diurnal space around the centre of the sun to the nocturnal space.

For in all these things, the nature of man, the observer creature and future dweller on the Earth, was taken up among the archetypal causes—as being one who was going to reckon the magnitude of the solar body and contemplate the differences of day and night.

[552] *But if this had been sought, it seems that it would also have been obtained. But you yourself admit that those ratios have been disturbed, since on account of those further increments from the sun the 360 days have become 365¼ and thus the diurnal journeys have become shorter.*

1. It cannot be said that this was sought simply, but merely in the adjustment of the principle of internal movement in the Earth: and in that manner it was obtained. 2. But even if in this secondary movement the concourse of causes disturbs the number instituted; still this disturbance was not so great but that during the months of November and January this magnitude is obtained: because then the magnitude of the diurnal movement of the sun is 1° or twice 30'. And long before, even if there were no such disturbance, the magnitude of the diurnal movement of the sun would have been such only twice during the year, on account of the necessary irregularity of the apparent movement of the sun.

How does the sun strengthen the motor virtue of the Earth by increasing the speed of the diurnal revolution of the Earth?

It is very probable that this takes place by the mediation of the light of the sun [*mediante solis lumine*], which the sun pours upon the Earth, through the illumination of the Earth's hemisphere. For because the physical additosubtractive difference [*aequatio*] of time demands unequal daily revolutions of the Earth, according as its distances from the sun vary; certainly at a short interval the illumination is strong as coming from a denser light; and weaker at a long distance, as from a thinner and thus lesser light; and that occurs—as regards the one dimension of longitude, into which the movement proceeds—in the very ratio of the distances. And so the amount of light which there is at any given time is fitted, by means of the distances, for distributing this acceleration throughout the year.

[553] *What are the effects of the daily revolution of the Earth, and, in general, of the revolutions of the primary bodies around their axes?*

Two: the first, which is proper to the Earth, is that for us who dwell on the Earth, all the stars in the heavens, both the fixed and the wandering, and so also the sun and the moon are seen every day to rise in the east and set in the west; although with respect to this diurnal movement they remain fixed in their places. I have treated of this deceptive appearance in the first three books on the doctrine of the sphere. The other effect, which is physical and most true and is common to all the primary bodies, and hence to the sun, is that the primary planets, by means of the form departed from their body set in revolution, move their secondary planets—as the Earth, the moon—and cause the secondary planets to move in the same direction but more slowly, and as if left behind.

By what arguments is it made probable that the primary planets share their own movements around themselves with the secondary planets, and especially the Earth with the moon?

The moon and Earth give the first evidence. For, just as above, from the fact that the planets, on drawing near to the sun, are borne more speedily, we reasoned that the sun by means of the form from its body, *i.e.*, its form set in rotation, moves the planets around itself in the same direction; so also, because we find that (1) in so far as the moon draws nearer to the Earth—but not to the sun—so much the more speedily does it move around the Earth, and (2) in the same direction in which the Earth revolves around its axis; it is with the greatest probability that we derive that movement of the moon from the whirling of the Earth; and that is all the more probable, because (3) there is also the correspondence that, just as the rotation of the sun around its axis is shorter than the shortest period of Mercury, so too the Earth rotates approximately thirty times, before the moon has one restitution. For if the moon revolved more quickly than the Earth, its movement could not wholly come [554] from the rotation of the Earth. (4) But belief in this thing is confirmed by the comparison of the four satellites of Jupiter and Jupiter with the six planets and the sun. For even if in the case of the body of Jupiter we do not have the evidence as to whether it rotates around its axis, which we do have in the case of the terrestrial body and the solar body in particular, that is to say, evidence from sense-perception. But sense-perception testifies that exactly as it is with the six planets around the sun, so too is the case with the four satellites of Jupiter: in such fashion that the farther any satellite can digress from Jupiter, the slowlier does it make its return around the body of Jupiter. And that indeed does not occur in the same ratio but in a greater, that is, in the ratio of the 3/2th power of the distance of each planet from Jupiter: and that is exactly the same as the ratio which we found above among the six planets. For Marius in his *World of Jupiter* reports that the distances of the four Jovial satellites from Jupiter are as follows: 3, 5, 8, 13 (or 14 according to Galileo); just as if their small spheres were separated by the three rhomboidal solids. (I) The rhomboidal dodecahedron between the inmost spheres, whose intervals are 3 and 5. (II) The rhomboidal triacontahedron (folium 464) between the middle spheres, whose intervals are

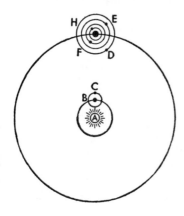

5 and 8, and (III) the cube, which is not truly rhomboidal, but is a beginning of the rhomboidal solids, between the extremes, which have the intervals 8 and 13 (or 14). Now the same Marius reports the periodic times as follows: 1 day 18½ hours; 3 days 13⅓ hours; 7 days 2 hours; 16 days 18 hours; everywhere the ratio is greater than double, and is accordingly greater than the ratio of the intervals 3, 5, 8, 13, or 14, but smaller than that of the squares, which constitute the ratios of the 2nd powers of the intervals, [555] namely 9, 25, 64, 169 or 196; just as also the ³⁄₂th powers are greater than the 1st powers, but smaller than the 2nd powers.

Therefore since the agreement of the Jovial satellites with the six primary planets is so exact, not only did we rightly infer above from this that the body of Jupiter turns around its axis like the sun, so that the proportion holds for all its members; but here already, over and above that, we are confirming not improperly the general statement that this rotation of the primary bodies around their axes is the cause of the circuit of the secondary bodies around the primary bodies. That (5) is so much the more probable because we see that, just as the sun is greater than all the planets which it moves, so too the Earth is much greater than its moon, and Jupiter than its satellites, and for that reason, like the sun, fit for moving them. The remaining probabilities have to do with the moon again. For (6) that the bodies of the moon and earth are akin is taught us by the telescope, which reveals signs in the moon of mountains and of seas, such as they are in our globe of the Earth. Even Aristotle, elsewhere a very sharp defender of the fifth essence of the heavens, recognized this kinship, and, according to Averroes, he said that the moon seemed to be a certain ethereal Earth. I am silent as to Plutarch and the other philosophers in Macrobius.

Accordingly, just as the kinship of their bodies makes the loadstone attract loadstone or iron; so also in the case of the moon it is not unbelievable that she should be moved by the terrestrial body which is akin, although neither in that case nor in this case is there any contact between the bodies. And furthermore (7) why is it surprising that the moon should be moved by the Earth, since we see that in turn the moon also, by its passing above the vertices of places, causes the ebb and flow of the ocean on the Earth? Is not this a clear enough evidence of the sharing of movements by these two bodies? Finally (8) the same thing is confirmed by the remaining part of the proportion: the sun and Earth wheel around their axes, as experience makes us certain—in the case of the sun, of itself [per se], in the case of the moon, only in Copernicus—namely in order that by this rotation they may give movement to the planets placed around them—the sun to the six primary planets, and the Earth to the moon: that the moon in turn does not wheel around the axis [556] of its own body is argued by the spots. But why is this so? If not because no further planet is seen to go around the moon. Accordingly the moon has no planet to which it gives movement by the rotation of its body. Accordingly, in the moon, the rotation was left out, as being superfluous.

If these eight arguments are not of use as taken singly, they will be of service as taken together.

But does it not seem absurd that the Earth, which lacks light, should be compared as an equal with the sun, the source of light? For does not this quality make the motor force of the sun more probable?

Even if the light [*lumen*] of the sun works itself up in supplying movement, nevertheless the body of the sun is not potent in motor force on account of the light alone. For there is nothing to prevent two, so to speak, subjects of motor virtue from being found together in the sun—light and the bodily affection of magnetism—and only the latter of them being present in the Earth; just because the Earth moves only one planet and that a very ignoble—as one of the secondary planets. Nor does the magnetic virtue of the Earth alone move it without any assistance, as we hear; nor does the Earth have this force wholly *from* itself, though the force is *in* the Earth; but, the Earth seems to have partly drawn off this force as by a certain canal by the continuation of the line from the sun into itself and especially in the illumination of its body, and it seems to have turned aside this force into a new source, namely into its own body—as was said a little while before, and will be said more clearly below.

The rotation of the Earth keeps to the equatorial circle; but the movement of the moon, to the zodiac, which has a great declination from the equator. Therefore it is not probable, is it, that the movement of the moon comes from the rotation of the Earth?

This does not bother us any more in the case of the moon than in that of the other planets. For even if they have a declination towards certain regions of their own [557] and, so to speak, hold the rudder in their hands and turn at their judgement and sail sideways or towards the banks of the river, none the less they are seized by the force of the common whirlpool of motion flowing out from the sun; and so they have the movement of the common river to thank even for the distinct movement of their own, just as the moon has the direct movement of the Earth along the equator to thank for its own oblique movement through the zodiac.

Why therefore does the moon keep its whole route along the zodiac rather than along the equator?

Because in addition to the proper circuit of the moon around the terrestrial globe—concerning which up to now—the whole heaven of the moon is also moved in the movement shared with the centre of the earth along the zodiac around the sun, like the other planets. And it comes about from the composition [of these movements] that with respect to the centre of the sun the moon always holds to a direct course eastward not only at the time when the sun and the Earth, with their distances from the moon overlapping, hurry the full moon into the same region, but also at the time when the sun impels the dark or void moon forward, while the Earth—with respect to the centre of the sun—impels it backward. For this impulse from the Earth is still much less than that from the sun: wherefore it diminishes the movement's bearing eastward, but does not wholly absorb it, much less turn it in the contrary direction. See the diagram of this composite movement of the moon in the *Commentaries of Mars,* folium 149.

Therefore since that flow of the solar form advancing along the zodiac is greater, and is other than the terrestrial form, which, proceeding along the equator, is less; and moreover since when the moon and sun are in conjunction, by reason of the speed and the region of rising and setting, in the space of the world more is in obedience to the sun than to the Earth: I believe that hence it comes about that by reasons of the latitudinal regions too, the moon is more obedient to the solar form as the stronger—as its whole heaven around the sun, so also its body around the Earth—and is compelled to advance along the zodiac [558] or to regulate according to zodiac its own orbit around the Earth.

From this does not anomaly arise in the lunar movement, if the moon at the tropic points advances according to the lead of the terrestrial form, because the zodiac and the equator are parallel at those points, but at the equinoctial points it crosses obliquely this form from the terrestrial body?

Again, in order to solve this objection, I make the same answer as in the case of the latitudes. Namely, that the form from the terrestrial body is very strong in its midpart under the equator, but is weaker to the sides of the equator; because even at the source, namely in the terrestrial globe, the circles parallel to the equator, as being smaller, are set in motion more slowly than the equator, the great circle. Accordingly a balancing takes place: so that where the moon experiences a strong motor form, there it does not comply with the whole form but goes off crosswise—and where it complies with the whole form and is utterly obedient to it, there the moon experiences it weak. Notwithstanding, I do not proclaim anything concerning the balancing in all its respects, since lunar observations still disagree in very small amounts from any given calculations, and since it is uncertain to what that discrepancy should be referred.

How too can the moon be borne around the sun in an annual movement, but the four satellites around Jupiter in a common duodecennial movement, so that meanwhile neither does the moon abandon or let go of the Earth nor the four Jovial satellites Jupiter, if the moon is not bound by a solid sphere to the Earth, and they to Jupiter?

Now the secondary planets are borne around the sun by the same virtue of the solar form, whereby also their primary planets, the Earth and Jupiter, are borne; but they would revolve more speedily than their primary planets, inasmuch as they are readier for movement by reason of their density, bulk, or weight, if they were not held back [559] and laid hold of by the Earth and Jupiter by means of a magnetic force similar to that with which the sun too is endowed. But this prehensive force, as was said above concerning the planets too, is determined by the contrary virtues of the approach and withdrawal of the moon from the Earth; for as the Earth revolves around its axis, by this laying hold it makes the moon revolve but meanwhile changes the region of its own body with respect to which approach and withdrawal takes place. See the diagram on page 899. Imagine the friendly region (*plagam amicam*) of the lunar globe to be turned towards the Earth and not to be changed about with the contrary region. Imagine also that the Earth does not rotate around its axis, but is nevertheless borne around the sun: in this case the moon will run the same course as the Earth, and meanwhile it will be attracted by the Earth, until it comes into contact with it. In turn imagine the same thing concerning the unfriendly region: in this case the moon will flee the Earth until it gets outside of the sphere of magnetic virtue of the Earth: then it will wholly give itself up to seizure by the sun alone, and thus will wander completely away from the Earth.

You have said that the middle circle of the Earth is slightly less than sixty times narrower than the sphere of the moon. But this same circle of the Earth is only thirty times faster than the moon because the moon returns in 29½ days. Therefore the circle of the Earth is slower than the centre of the moon around the Earth in the ratio of two to one. How then does a body, which proceeds more slowly, give to the moon a movement twice as great and twice as fast as its own movement?

This objection does not apply uniquely to lunar movement, but generally to all the planets; and there is nothing absurd in it. For the solar and terrestrial bodies do not move [things] by contact, but by the spreading out or unfolding of their forms into every orbit of the movable body. Now in so far as the form from the terrestrial body flows out through space, it turns with the Earth, its source, in the same time of 24 hours, since nevertheless in that place where it lays hold of the moon, it is of the same amplitude as the sphere of the moon. [560] Therefore that form, sixty times wider than the Earth, passes through the total orbit of the moon thirty times in one month, although within the same interval the moon, following after the form from the Earth, has only one periodic return. And so it remains probable that the movement of that form from the terrestrial body moves the moon; but in such a way nevertheless that the inertia of the lunar body overcomes diurnally about 29 parts of the expanse of virtue and is overcome with respect to not more than the thirtieth.

Why do you set down that the sun concurs with the motor form of the Earth even in that movement whereby the moon revolves around the Earth?

1. Because Tycho Brahe found that the mean movement of the moon—that is, without the anomaly which exists in all planets on account of the eccentricity of the orbit—[561] still has an anomaly or is irregular. For the moon is always speedier in the syzygies, as here in *CD* and *GH*, and slower in the quadratures *EF* and *IK* than the ratio of eccentricity accounts for—whether in either case it is in the apogee or in the perigee or in any other position on its eccentric circle; and—if we are to insist firmly upon Tycho's hypothesis for the said variation—the moon is exactly as much faster in the syzygies as it is slower in the quadratures.

But the form itself from the Earth—the form set in rotation and to be understood by the circle *DFHK*—is of uniform speed all around, as much

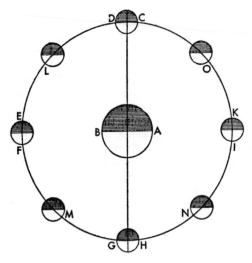

in those parts which are turning at the syzygies D and H as in those parts which are turning at the quadratures F and K—that is to say, for one and the same interval between the moon and the Earth. Accordingly other causes of movement which may be adjusted to the phases of the moon must be added to this motor form. But the phases of the moon are caused by the sun. Therefore the sun assists the movement of the moon around the Earth.

2. Belief in this concurrence of the sun is strengthened by the fact that before this, on pages 917 and 918, the same sun was summoned in order to aid the speed of the Earth in revolving, by the illumination of the [terrestrial globe] —here represented by the middle circle AB. For from this, as in the example of the Earth, we first understood that even in the light of the sun there is present a power of quickening movement. Then from the same thing we weave a necessary argument even for the moon. For if, according as $DFHK$ the form from the terrestrial body AB is set in whirling motion, it moves the moon; but the sun accelerates this whirling motion: therefore the sun will accelerate the moon too by means of the Earth and the acceleration of its form.

Then this illumination is not disposed in one way when the moon is turning in the quadratures F and K, and in another way when the moon is turning in the syzygies D and H?

By no means. For in both cases the halves of the globes are illuminated, both of the Earth AB, which gives movement, and of the moon CD or GH, to which movement is given. Rather, it has already been said that the speed of the Earth from this illumination [562] is equal at both times.

Whence then does this disparity of effect come to this accessory cause, so that it accelerates the movement of the moon very much in the syzygies D and H but not at all in the quadratures F and K? And what on the contrary slows the movement of the moon in the quadratures F and K?

No part of celestial physics was more difficult to explain than this. And in order to straighten things out, where possible, we must make use of the diagram on this page.

Accordingly you remember that all the circles which bound the illumination of the lunar globe, as CD and GH, were parts of the same number of spherical surfaces, into which the light coming from the sun as from a centre is spread out; while the circle $DFHK$ represents the form of the terrestrial body AB at the center of its position, and this form moves the moon. You see that in the syzygies D and H the form from the light CD and the form from the terrestrial body $OCDL$ are joined to one another by contact (*invicem applicari per contactum*), and that at L, M, N, and O they cut one another at oblique angles, so that the joining (*applicatio*) is more incomplete. But in the quadratures EF and IK the cutting is at right angles; therefore there is clearly no joining, since the section of the moon stretches towards the centre of the Earth and a mere point on the circle NIO corresponds to it.

Therefore, since no other cause for the acceleration in the syzygies is apparent, we shall have to set down that a faculty strengthening the terrestrial motor form ODL is present in the light CD separately, not now in so far as the source of the light, *i.e.*, the solar body, is rotated—this modification given by the movement was valid above, since we were speaking of the forms of the solar and terrestrial bodies without any reference to light—but *qua* light: indeed, in accordance with the true and somehow essential configuration of light. Then if we set down that this form from the terrestrial body is strengthened through the modes of its joining to the circles (*orbes*) of light; we shall have at hand the cause and the measure of there being a very strong acceleration in the syzygies CD and GH, but none in the quadratures.

[563] But since by $DFHK$ is represented not merely the form of the terrestrial body as agent or mover, but also the very orbit of the moon as of the patient or thing moved—although at that time the Earth must not be placed at the centre of the circle but near by—we must conceive further either that at the syzygies CD and GH the lunar body is better disposed towards movement according to the diffusion or the surface of the light than in EF and IK, where the moon cuts the circles of diffusion crosswise, or that the road of the moon in D and H is made as it were slippery, but rough in F and K—as on a table with grooves running crosswise in the wood. And that is not very absurd. For since the power of strengthening the movement is present in the light, as it was laid down; surely where one dimension of the light stretches, it is likely that the passage is easy.

Furthermore, as regards the effect, the same thing is said by him who says that the moon is accelerated in D and H and slowed up in F and K and in both cases in the simple ratio which these joinings (*applicationes*) produce, and by him who says that the moon is very greatly accelerated in D and H and none at all in F and K, but in the ratio of the squares of that which results from the joinings posited here.

Unless anyone prefers to ascribe this twofold efficacy of the light to the two dimensions of the surface of the light; so that, although any immaterial forms of bodies are diffused no less than light is diffused both in longitude and in latitude; nevertheless those forms up to now have been effectual only with reference to longitude; but this [light] is effectual with reference both to longitude and latitude: on account of the fact that the form moves as moved; but it is moved only in longitude; while light strengthens as light, *i.e.*, according as it possesses its own density, both in longitude and in latitude.

Why do you attribute the force to strengthen the motor cause to the light separately and outside of its reference to the rotation of its source?

Because in so far as the form from the rotating source moves [the moon], it always moves it eastward in the direction $CIDL$; and in the beginning of this section we already finished with its effect [564] in moving the moon. But this force from the light is potent in accelerating the moon even westward in the direction *MNH*, with reference to the centre of the sun, at that time namely when the moon appears to us to be void of light or in conjunction with the sun. Therefore the light does not by itself bring movement and direction together (*conciliat motui plagam*), but by means of the acceleration of the form *MHN*.

If this force is present in light, greater force will be present in the denser light around GH, *as being in the neighbourhood of the sun; a lesser in the more scattered light, around the full moon in* CD, *since it is farther away from the sun by a thirtieth part of the distance: therefore the new moon will be speedier than the full, other things being equal.*

The more perfect joining balances the weakness of the light *CD*, as *CD* is of a more regular concavity than *GH*. Therefore since the strengthening takes place through the joining of the forms: in the full moon the more scattered light, joined more perfectly, accomplishes as much as in the new moon the denser light, joined more imperfectly, does. Now the intervals between the moon and the sun which measure out density to the light and which measure out the curvature to the circles *DC* and *GH* are the same: wherefore the density is perfectly balanced in longitude by the curves *CD* and *GH*. But the effect of the light on one side is balanced by the diversity of the joining on the other. For even though *CD* and *GH* are equally curved; nevertheless in the first case the convex *OCDL* winds into the concave *CD*; in the second case the convex *MGHN* is turned towards *GH* the form from light, which is convex towards the Earth.

If that addition of 133° to the 12 restitutions (synodos) *in the sidereal year comes from the acceleration of the lunar movement in the syzygies, the magnitude of the acceleration must correspond too.*

Indeed in Tycho Brahe the movement [565] of the moon in the syzygies is accelerated only 1'26" per 1°, and is slowed 1'26" per 1° in the quadratures; wherefore if the slowing is effaced by the twofold acceleration, the greatest acceleration of the syzygies will be 2'52". Wherefore if the sines squared of all the 90° bring their small portions into one sum, we shall have as the aggregate 2°9'; therefore in the sidereal year 106°22' but not 132°45'.

But in the first place the magnitude of the greatest variation is not very certain in Tycho, who exhibits it as 40½' at 45°; and so, if the variation is set down as 51', we equal the aforesaid sum—3'34"40''' being taken as the acceleration of 1°, or in Tycho's formulation 1'47"20''', and an equal slowing up at 90° or in the quadratures; and thus in one quadrant the sum of 2°41' is added up, and that sum will acquire great probability below, when we deal with the causes of the irregularities. Then if in particular we keep Tycho's small magnitude, at 45°; the preceding and the subsequent magnitudes, distributed in another formulation than Tycho's, could give the desired sum; or else there are hidden from us very minute causes, which take away something from the 133° in the treatment of the variation.

Then in what proportion do you think the monthly movement of the moon around the Earth should be distributed among those two causes, namely the form of the terrestrial body and the circle of illumination of the bodies?

We see that while the Earth revolves around its axis approximately 29½ times, in the meantime the moon returns around the Earth once, namely from sun to sun. So it happens that in one year or 365 days 6 hours 9 minutes 26 seconds the moon returns twelve times and adds on more than one third of the thirteenth revolution, that is, 132°¼′. Therefore it is likely that the density of matter in the lunar body is so proportioned to the archetypal degree [566] of strength in the form from the terrestrial body, that unless illumination aided the daily revolution of the Earth and also by means of this [revolution] aided the progress of the moon, the moon itself by reason of the simple motor virtue of the Earth would return slightly more slowly, that is to say, exactly twelve times. With this laid down, it follows that those remaining and as it were supernumerary 132¼° of the incomplete thirteenth revolution must be attributed to the other motor cause, namely the illumination.

Then you evaluate the density of the lunar body as proportioned to twelve lunar revolutions in one year: what will you say is the archetypal cause of this number?

The cause seems to be composite of geometrical beauty and of the office of this planet in the world: as follows: For the moon is a secondary planet assigned to the Earth, and it keeps to its own private course around the Earth. But 360 revolutions were allotted to the Earth, while the centre of the Earth makes one return around the sun. Then, just as among the upper planets, the sphere of the moon had to be a mean proportional between the body of the Earth and the sphere wherein the centre of the Earth really revolves, but the sun apparently; so also the revolutions of the moon had to be more than one but fewer than 360. And indeed the mean proportional between 1 and 361 is 19; but because the number 361 is not 360 and because 19 does not have any beauty either geometrical or harmonical; then the two numbers nearest to 19, which when multiplied together, give 360 and are the most beautiful geometrically and harmonically, should be chosen. Now the nearest numbers which give 360 are 18 and 20, because the first is smaller by unity alone, and the second greater than 19 by unity alone. But there is no demonstration of a figure of 18 sides. The numbers following nearest are 15 and 24, which also give 360. Now there are geometrical demonstrations of them, but rather worthless ones; nor do they give any outstanding ratio, [567] but only the ratio between 5 and 8; nor are they the most excellent and first of all in harmonics. But the numbers 12 and 30—for there are no others nearer which give 360—excel in all ways: both geometrically, as being generated from the first figures inscribed in the circle, and harmonically, because all harmonies are represented by these two divisions of the string. Then of the numbers which, when multiplied together, give 360, there are none more beautiful.

Furthermore, a number [not] less than 12 and not greater than 30 was due to the revolutions of the moon; because the sphere of the moon exhibits an image of the sphere of the sun, it was fitting too that, just as the year, which is the periodic time of the sun, is divided into 360, a great multitude, so also the month, which is the periodic time of the moon, should be allotted parts or days greater in number than all the months in the year; and that the multitude should increase with the progress, if in the first place the year, the long time, were divided into 12 months, the big parts, and thence the month, the short time, into 30 days, the little parts; for multitude is becoming to small things.

And it would not have been of the same beauty, if there had been thirty months in the year, and each month had been of twelve days.

How do you make it probable that the increase in the annual revolutions of the Earth above the number 360 and this addition to the annual movement of the moon beyond the twelve monthly revolutions of the moon come from the same cause?

The very reasons of this philosophy bear witness to this thing; so that, because the daily rotation of the terrestrial globe moves the moon, the greater in number and speedier the rotations are, they move the moon with greater speed, and make it return more often. And especially so does the comparison of the number of days in the solar year—365 days 6 hours and a little more—with the archetypal number 360 and with the number of days in the lunar year—354 days and a little less than 9 hours.

[568] For since 360 days in the year—making the moon revolve 12 times—should have come from the archetype, but they have become 365 by the intervention of the other cause; accordingly all the revolutions have become faster in the ratio of 360 to 365, and in that ratio stronger with respect to moving the moon. But at the same time they have become greater in number, *i.e.*, 365. Therefore the faculty of the 360 archetypal revolutions should be evaluated at the number 360. But the faculty of these actual 365 revolutions should be evaluated not at the number 365, because these revolutions are faster, but at the number which is a third proportional, namely 370°36′50″—if we attend to the minutiae. But if the faculty stamped with the number 360 would have moved the moon so that it would have completed 12 returns with respect to the sun and would have ended the last of those returns at its initial position beneath the fixed stars: therefore, in the same proportion, the faculty evaluated at the number 371[1] will make the moon outrun the sun 12 times and go 127°10′ beyond its initial position: and because, after the 360 days which were in the archetype have been completed, the sun is still 5°10′ distant from its initial position beneath the fixed stars, and because the circle, which in the archetype had been divided among the 12 positions of the full moons, has been narrowed by that interval, accordingly the addition of these 5°10′ to those 127°10′ makes 132°20′. See how near this reasoning comes to the truth in the astronomical tables; for they give the excess of the moon in the sidereal year as 132°45′, only 25′ more.

We will also infer the same thing from the days of the lunar year, as follows: The motor faculty of the 360 revolutions of the Earth would have given the moon its twelfth return with respect to the sun and to its initial position; therefore the faculty of revolutions fewer in number but so much the stronger will accomplish just as much. Accordingly as 365 revolutions is to 360, so the faculty of the 360 archetypal days is to the faculty of the 354 actual days and 19 hours, 33 minutes. Therefore so many terrestrial revolutions, now become more intense, would have given the moon its twelfth return with respect to the sun, if only the intervals between two syzygies had not been contracted by the increase in the number of the revolutions. But because, with the insertion of the supernumerary days in the year, the 360th day as [569] archetypal breaks up the measure of contraction of the zodiac, from which 5°6′41″ are due proportionally to the length of the lunar year; accordingly the moon too is relieved [*sublevatur*] of the same number of degrees; so that even when those degrees have not been

[1]To be precise: in the actual computation Kepler uses the number 370°36′50″.

traversed in the expanse of the world, nevertheless the moon makes its twelfth return with respect to the sun. Now those degrees are equivalent to 10 hours 4 minutes. And when 10 hours 4 minutes have been subtracted from 19 hours 33 minutes which were found, 9 hours 29 minutes remain in addition to the 354 days. Instead of these 9 hours 29 minutes the astronomical tables give 8 hours 49 minutes, so that less than an hour is missing. And that small difference can be assigned to other minute circumstances. Meanwhile it has been proved exactly enough in both ways that this straying from the whole and beautiful numbers is due to the concurrence of the causes of the lunar movement. And the reason is clear why 360 is approximately a mean proportional between the lengths of the lunar year and the solar sidereal year.

PART III

On the Real and True Irregularity of the Planets and its Causes

From what do the planets have that name which signifies wanderers in this language?

From the manifold variety of their proper movements. For if you follow the judgment of the eyes, that variety has no law, no determined circle, no definite time, if a comparison is made with the fixed stars.

In how many ways do the planets seem to wander?

In three ways: (1) In the longitude of the sphere of the fixed stars, which we said extends along the ecliptic. (2) In latitude [570] or to the two sides of the ecliptic, towards its poles. (3) In altitude, *i.e.*, in the straight line stretching from the centre of vision into the depth of the ether. Nevertheless this variety is not uncovered by the eyes alone; but reasoning from the diverse apparent magnitude of the bodies and the arcs assents to it.

What must be held concerning these wanderings of the planets? Do they really wander in so many various ways, or is sight merely deceived?

Even though that movement is not wholly such as meets the eyes, it is present in the planetary bodies themselves; but much deception of sight winds its way in here. Nevertheless when these deceptions have been removed by the mind, some irregularity of movements still remains and is really present in all the planets.

Then what is that true movement of the planets through their surroundings?

It is constant with respect to the whole periods; and proceeds around the sun, the centre of the world, always eastward towards the signs which follow. It never sticks in one place, as though stationary, and much less does it ever retrograde. But nevertheless it is of irregular speed in its parts; and it makes the planet in one fixed part of its circuit digress rather far from the sun, and in the opposite part come very near to the sun: and so the farther it digresses, the slower it is; and the nearer it approaches, the faster it is. Finally, in one part of the circle it departs from the ecliptic to the north, and in the other, to the south. And so the planet is left with a real irregularity and one which is threefold, too: in longitude, in latitude and in altitude. The astronomers prove that by suitable evidence—concerning which in Book vi.

1. THE CAUSES OF THE TRUE IRREGULARITIES

[571] *State what the ancients thought about the causes of this irregularity.*

The ancients wished it to be the office of the astronomer to bring forward such causes of this apparent irregularity as would bear witness that the true movement of the planet or spheres is most regular, most equal, and most constant, and also of the most simple figure, that is, exactly circular. And they judged that you should not listen to him who laid down that there was actually any irregularity at all in the real movements of these bodies.

Do you judge that this axiom should be kept?

I make a threefold answer: I. That the movements of the planets are regular, that is, ordered and described according to a fixed and immutable law is beyond controversy. For if this were not the case, astronomy would not exist, nor could the celestial movements be predicted. II. Therefore it follows that there is some conformity between the whole periods. For that law, of which I have spoken, is one and everlasting; the circuits or traversings of the celestial course are numberless. But if they all have the same law and rule, then all the circuits are similar to one another and equal in the passage of time.

III. But it has not yet been granted that the movement is really regular even in the diverse parts of any given circuit. (1) For astronomy bears witness that, if with our mind we remove all deceptions of sight from that confused appearance of the planetary motion, the planet is left with such a circuit that in its different parts, which are really equal, the speed of the planet is irregular—just as there is apparent inequality in the angles at the sun which are equal with respect to time. And Ptolemy himself, by setting up different centres in accordance with the rule of movement of eccentrics and epicycles, makes those circles of his to move more swiftly at one time, and more slowly at another. (2) [572] Finally, astronomy, if handled with the right subtlety, bears witness in this case that the routes or single circuits of the planets are not arranged exactly in a perfect circle but are ellipses.

But by what arguments did the ancients establish their opinion which is the opposite of yours?

By four arguments in especial: (1) From the nature of movable bodies. (2) From the nature of the motor virtue. (3) From the nature of the place in which the movement occurs. (4) From the perfection of the circle.

Will you state their argument from the nature of bodies?

They reasoned that those bodies are not composed of the elements, and so neither generation or corruption nor alteration has any rights over them. The experience of all the ages bears witness to this: for the bodies are always viewed as the same and are not found to have changed at all in mass or in number or in form. But the movements of the bodies made up of elements are for this very reason various and inconstant, because the elements are variously mixed in the constitution of the bodies and are at war with one another within the mixed bodies. Therefore in the celestial bodies, where there is no such mixture and no war of the elements as in mixed bodies, there is also no place for turbulence, none for irregularity.

What answer do you judge should be made to this argument?

If the argument is speaking of disordered turbulence of movements, there is none such in the heavens: there are no celestial disturbances as in thunderstorms,

Flame and drops of water at war with one another

because the composition of the bodies of the world is of a very different family. But if the argument is in opposition to every regular irregularity also; [573] then not every irregularity, certainly not that regular intensification and remission of movements, comes from the war of the elements mixed together in the moved bodies, nor from the bodies being mutable. For some irregularity arises just because they are bodies, bodies which are moved and which give movement too, and because they are made up of their own matter, their own magnitude, and their own figure both inwardly and outwardly, and in accordance with their magnitudes and figures they are endowed with their natural power too. And in accordance with their natural power they are less movable at a distance than at near-by, where the faculties of mover and moved are in agreement rather than at war. Thus by one part of its body the loadstone attracts iron, and by the other repels iron; not in either case on account of any mixture of elements, but on account of the inward rectilinear configuration, in accordance with which the loadstone has an inborn virtue. Thus the same loadstone attracts more strongly iron when near-by than when farther away—not that when the loadstone is nearer, it has more of fire or Earth, but because its virtue is weakened with its distance. Nevertheless the celestial bodies—*i.e.*, the bodies of the world—remain everlasting and immutable as regards their total masses [*moles*]. For the changes which come about on their surfaces can bring on nothing sufficient to disturb [*nullum momentum ad turbandos*] the movements of the total masses [*molium*]. And upon this everlastingness of the whole globes and upon the fact that in the world there is nothing disordered which impedes their movements, there depend this regularity of the circlings, and the everlasting similarity, and the constant regularity, with respect to the whole cycles, of the irregularity in the single parts.

Will you review the second argument of the ancients taken from the mover cause?

They said, the motor virtues of the celestial bodies are of the most simple substance; they are minds divine and most pure, which do unceasingly what they do; they are everlastingly similar; they employ a most equal struggle of forces, and they are never tired, because they feel no labour. [574] And so there is no reason why they should move their globes differently at different times. And accordingly even the figures of the movements, on account of the very nature of the minds, are most perfect circles.

What do you oppose to this?

Even though the motor virtue is neither some god nor a mind, nevertheless we must grant what the argument intends—chiefly too in the case of that motor cause which a truer philosophy brings in, namely in the case of the natural power of the bodies, because wherever and in so far as such a power is alone, it moves [a body] most regularly and in a perfect circle, and does that by the sole necessity of effort and by the everlasting simplicity of its essence. That is the case in the rotation of the solar body and in that of the Earth too in especial; for this rotation comes from one sole motor cause: whether it be a quality of the body or a sprout of the soul born with the body. For the axis with its two opposite poles stays fixed; but the body revolves around the axis most regularly and in a circle. This would be the case still, if any planetary globe were always at the same distance from the sun. For it would be carried by the sun with utmost regularity in a perfect circle, by means of the immaterial form released from the solar body set in a very regular movement of rotation. And by reason of this same very regular movement even that form from the body revolves in the amplitude of the expanse of the world, like a swift whirlpool.

But although so far we have granted the argument of the ancients, nevertheless regularity of movements in every respect does not yet follow from this. For not only do the motor virtue and the movable body come together in the movements, but also the inward rectilinear configuration of the movable body; and in proportion to its diversity of posture in relation to the sun, this configuration is affected in diverse ways in the movement: in one region it is repelled, in another it is attracted towards the inside. The axis of the magnetic movable body comes in, and so does being at rest in the parallel posture; and from that inward repose and from that revolution coming from outside there results that change of posture [575] of the parts of the planet in relation to the sun. Finally there advenes the interval between the sun and the planet, and this interval varies with the attraction and repulsion. But when the interval has been changed and the planet comes into a denser or more rarefied virtue, then its movement too necessarily suffers intensification or remission, and the figure of its route becomes elliptical. So with reference to the concourse of so many required things, the virtue moving the planet cannot be called simple, because it moves by means of different degrees of its form.

What was the ancients' argument from place?

They considered that the region of the elements was around the centre of the world; the heavens at the surface. Therefore to the bodies made up of elements belongs rectilinear movement, which has a beginning and an end and which, disbursed according to the contrary principles of heaviness and lightness, brings any of those bodies back into its own place; and hence in proportion to different nearnesses to the natural place, or mark, there are different speeds, and finally pure rest. But the celestial bodies move everlastingly in the circular expanse of the world: and that argues that they are neither heavy nor light, and that they are not moved for the sake of rest or for the sake of occupying a place—for they are always circling in their place—but that accordingly they are moved only in order to be moved; and so their movement must be regular, and the form of their movement must be other than rectilinear, namely suited to an eternity of movement, that is, returning into itself.

What answer do you make to this third argument?

Not every irregularity of movements comes from heaviness and lightness, the properties of the elements; but some comes from the change of the distance too, as is clear in the case of the lever and the balance; and this cause produces intensification [576] and remission of movements, as has been explained so far. We must however remark that there is nevertheless some kinship between the principles of heaviness and lightness in the elements and the natural inertia of the planetary globe with respect to movement, but no irregularity of movement is explained by this kinship.

But as regards the figure of the movement, the argument concludes nothing more than we can grant, namely that the movement bends back into itself. And not only the circular but also the elliptical are of such a kind; and so the assumptions are not denied. For in truth bodies which revolve around their axes are moved only in order that by their everlasting motion they may obey some necessity of their own globe—some bodies indeed in order that they may carry the planets around themselves in everlasting circles.

State the fourth argument of the ancients which was taken from the circular figure.

They philosophized that of all movements which return into themselves the circular is the most simple and the most perfect and that something of straightness is mixed in with all the others, such as the oval and similar figures: accordingly this circular movement is most akin to the very simple nature of the bodies, to the motors, which are divine minds—for its beauty and perfection is somehow of the mind—and finally to the heavens, which have a spherical figure.

How must this be refuted?

To this I make answer as follows: Firstly, if the celestial movements were the work of mind, as the ancients believed, then the conclusion that the routes of the planets are perfectly circular would be plausible. For then the form of movement conceived by the mind would be to the virtue a rule and mark to which the movement would be referred. But the celestial movements are not the work of mind but of nature, that is, of the natural power of the bodies, or else a work of the soul acting uniformly in accordance with those bodily powers; [577] and that is not proved by anything more validly than by the observation of the astronomers, who, after rightfully removing the deceptions of sight, find that the elliptical figure of revolution is left in the real and very true movement of the planet; and the ellipse bears witness to the natural bodily power and to the emanation and magnitude of its form.

Then, even if we grant them their intelligences, nevertheless they do not yet obtain what they want, namely the complete perfection of the circle. For if it were a question only of the beauty of the circle, the circle would very rightly be decided upon by mind and would be suitable for any bodies whatsoever and especially the celestial, as bodies are partakers of magnitude, and the circle is the most beautiful magnitude. But because in addition to mind there was then need of natural and animal faculties also for the sake of movement; those faculties followed their own bent [*ingenium*], nor did they do everything from the dictate of mind, which they did not perceive, but they did many things from material necessity. So it is not surprising if those faculties, which are mingled together, could not attain perfection completely. The ancients themselves admit that the routes of the planets are eccentric, which seems to be a much greater deformity than the ellipse. And nevertheless they could not guard against this deformity by means of the province of those minds of theirs.

Now I have often reminded you that while I deny that the celestial movements are the work of mind; I am not at that moment speaking of the Creator's Mind, which all things indeed befit, whether circular or elliptical, whether administered and represented by minds or compelled by material necessity from the beginnings once laid down.

2. ON THE CAUSES OF IRREGULARITY IN LONGITUDE

[578] *Then what causes do you bring forward as to why, although all the routes of the primary planets are arranged around the sun, nevertheless the angles—in which as if from the centre of the sun, the different parts of the route of one planet are viewed— are not completed by the planet in proportional times?*

Two causes concur, the one optical, the other physical, and each of almost equal effect. The first cause is that the route of the planet is not described around the sun at an equal distance everywhere; but one part of it is near the sun, and the opposite part is so much the farther away from the sun. But of equal things, the near are viewed at a greater angle, and the far away, at a smaller; and of those which are viewed at an equal angle, the near are smaller, and the far away are greater.

The other cause is that the planet is really slower at its greater distance from the sun, and faster at its lesser.

Therefore if the two causes are made into one, it is quite clear that of two arcs which are equal to sight, the greater time belongs to the arc which is greater in itself, and a much greater time on account of the real slowness of the planet in that farther arc.

But could not one cause suffice, so that, because generally the orbit of the planet draws as far away from the sun on one side as it draws near on the other, we might make such a great distance that all this apparent irregularity might be explained merely by this unequal distance of the parts of the orbit?

Observations do not allow us to make the inequality of the distances as great as the inequality [579] of the time wherein the planet makes equal angles at the sun; but they bear witness that the inequality of the distances is sufficient to explain merely half of this irregularity: therefore the remainder comes from the real acceleration and slowing up of the planet.

What are the laws and the instances of this speed and slowness?

There is a genuine instance in the lever. For there, when the arms are in equilibrium, the ratio of the weights hanging from each arm is the inverse of the ratio of the arms. For a greater weight hung from the shorter arm makes a moment equal to the moment of the lesser weight which is hung from the longer arm. And so, as the short arm is to the long, so the weight on the longer arm is to the weight on the shorter arm. And if in our mind we remove the other arm, and if instead of the weight on it we conceive at the fulcrum an equal power to lift up the remaining arm with its weight; then it is apparent that this power at the fulcrum does not have so much might over a weight which is distant as it does over the same weight when near. So too astronomy bears witness concerning the planet that the sun does not have as much power to move it and to make it revolve when the planet is farther away from the sun in a straight line, as it does when the interval is decreased. And, in brief, if on the orbit of the planet you take arcs which are equally distant, the ratio between the distances of each arc from the sun is the same as the ratio of the times which the planet spends in those arcs. Thus let the centre of the sun or world be represented by the fulcrum of the lever, and its motor power by one arm and the weight on it—and we have already given the order to dissemble the arm and the weight, and mentally to reduce them to the fulcrum; but let the planet be represented by the weight on the remaining arm, and the interval between the sun and the planet, by the arm for that weight.

[580] Let AC be the lever, D and B the weights hanging from C and A, FE the fulcrum, and FEC and FEA right angles. As CE is to EA, so is the weight B on EA to the weight D on EC. Remove mentally EA, and let the power formed through EA by the weight B be the power of the fulcrum E; accordingly this power of the fulcrum E will keep the weight D, hung from C, in horizontal equilibrium, that is, so that FEC will be a right angle. But if

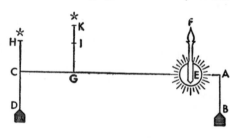

this same weight, pulled away from C, approaches as near as G, then the same power of E will have more might over this weight, and will lift it up above the line EC.

Now let E be not the fulcrum but the sun, and let D be the planet; and EC and EG the different distances of the planet from the sun. Accordingly observations bear witness that as EC is to EG, so is GK, the forward movement of the planet when nearer at G, to GI or CH, the forward movement of the planet when farther away at C.

Then do you attribute weight to the planet?

It was said in the above that we must consider that instead of weight there is that natural and material resistance or inertia with respect to leaving a place once occupied; and that this inertia snatches the planet as it were out of the hands of the rotating sun, so that the planet does not yield absolutely to that force which lays hold of it.

[581] *What is the reason why the sun does not lay hold of the planet with equal strength from far away and from near-by?*

The weakening of the form from the solar body is greater in a longer outflow than in a shorter; and although this weakening occurs in the ratio of the squares of the intervals, *i.e.*, both in longitude and in latitude, nevertheless it works only in the simple ratio: the reasons have been stated above.

3. The Causes of the Irregularity in Altitude

But what pushes the planet out into more distant spaces and leads it back towards the sun?

The same which lays hold of the planet, the sun, namely, by means of the virtue of the form which has flowed out from its body throughout all the spaces of the world. For repulsion and attraction are as it were certain elements of this laying hold. For repulsion and attraction take place according to the lines of virtue going out from the centre of the sun; and since these lines revolve along with the sun, it is necessary for the planet too which is repelled and attracted to follow these lines in proportion to their strength in relation to the resistance of the planetary body. So the contrary movements of repulsion and attraction somehow compose this laying hold.

Do you attribute to the simple body of the sun and to its immaterial form the operations of attraction and repulsion which are contrary and so not simple?

The natural action or ἐνέργεια of moving [582] the planetary body for the sake of assimilation or of bringing it back to its primal posture is one [in number]; but it seems to be diverse on account of the diversity of the object. For only in one region is the planetary body in concord with the solar body; in the other region it is discordant. But it belongs to the same simple work to embrace like things and to spit out unlike things. This opinion is strengthened by the case of magnets; for though they are not celestial bodies, nevertheless they do not have that biform virtue from the composition of elements but from a simple bodily form.

Therefore the planetary body itself will be composed of contrary parts?

No, indeed. For it follows only that the planetary globe has an inward configuration of straight lines or threads, like magnetic threads, which happen to be terminated in contrary regions; and in one of these regions, not on account of the body itself but on account of its posture in relation to the sun, there reigns friendship [*familiaritas*] with the sun; and in the other region, discord.

But isn't it unbelievable that the celestial bodies should be certain huge magnets?

Then read the philosophy of magnetism of the Englishman William Gilbert; for in that book, although the author did not believe that the Earth moved among the stars, nevertheless he attributes a magnetic nature to it, by very many arguments, and he teaches that its magnetic threads or filaments extend in straight lines from south to north. Therefore it is by no means absurd or incredible that any one of the primary planets should be what one of the primary planets, namely the Earth, is.

[583] *Granted that the planet has an inward rectilinear magnetic configuration: then what is it that makes the planet turn one region of its body after another towards the sun? It does not turn its threads about, does it?*

By no means: rather we should ask what it is which keeps the planetary body from removing its own magnetic axis from the posture which it has once taken with respect to the parts of the world, although nevertheless its body, like the terrestrial body, revolves around its axis and at the same time is moved out of its place and is transported in a circle around the sun. For out of this direction of the magnet towards the same region of the world during the whole circuit and out of the transportation of the body from place to place around the sun there is compounded, as out of two elements, the following effect: that the planetary globe changes the posture of its regions with respect to the sun. See the diagram on page 939.

What examples are there of this change?

Again there is a familiar example in the magnetic compass, namely where the iron needle has been magnetized. For into whatever place it may be transported, the compass needle always looks towards the north. And so if, carrying the compass, you march around some castle, then at one time the head and at another time the tail of the needle looks towards the castle, [584] because in every part of its circuit the head always looks towards the north.

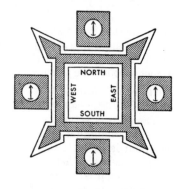

There was another astronomical example above in the third book, when we said that while the Earth revolves around the sun, the axis of terrestrial rotation remains in the same everlasting parallel posture—folium 248.

Then what causes do you assign for the pointing of the magnetic threads of the planetary body towards the same region of the world throughout the whole circuit of the planet?

The same as were indicated in Book I, folium 116[1], by which the axis of terrestrial rotation is made steadfast. For firstly the parallel posture of the threads manifests a certain sameness, which is rest rather than motion. The cause of that does not seem to be any natural power which is positive or active, but rather one which is privative of all movement. And so that natural inertia of matter with respect to movement seems to have an inward rectilinear configuration and to be extended in accordance with these threads or to be rendered stronger and more unconquerable by the condensation of parts in a straight line.

But if this is not probable: then let there be [two] distinct ἀδυναμίαι or powerlessnesses; the first belongs to all matter considered to be without inward configuration: and it enables the planet not to depart from its own place, unless it be drawn forth from the outside, namely by the sun. The second belongs to the planetary body according as it has a configuration of straight

---

[1]There are three possible causes. (1) Since the form of the turning is joined to the globe and is directed towards a definite region and not towards all indiscriminately, it follows with geometric necessity that the axis of this turning is directed constantly towards the lateral regions as long as the middle circle of the turning does not stray from its own region. (2) There is a private cause of motion, namely the natural inertia of the matter of the globe, due to the dragging of the axis, which necessitates force to turn it aside from its position; since, however, no power of motion is brought to bear against it, it remains at rest in its position. (3) There is an internal, positive, natural power in the rectilinear threads, which are parallel to the axis, to maintain them in their original position. For the power of movement nests in these threads of the entire globe; they are distributed circularly around the axis with equal strength on all sides and by means of them this power turns the body. Thus therefore in turn the axis, maintaining itself in its proper direction through a natural constancy fixes also the region of the turning, with the result that because of the inclination of the axis, the circle of the movement is also necessarily inclined.

threads on the inside, and by reason of it these threads are kept safe so that they are not deflected by the revolution of the body or moved out of their posture. Finally the philosophizers are free to determine whether what I have just spoken of is merely an ἀδυναμία or a δύναμις

You define this ἀδυναμία or δύναμις merely by the preservation of the posture. But what if something else were absent, and that δύναμις looked towards some other fixed parts of the starry heavens?

In Book I, page 116, when it was a question of the axis of terrestrial rotation which is similarly immovable, an answer was given as to why such a thing should not be thought of: namely because there is no reason why the axis should point towards some empty point in the heavens rather than towards some star, and why in this direction rather than in that; and because these planetary threads, no less than the axis of terrestrial rotation above, are found in the succession of ages to swerve slightly and thus to desert their original stars of reference [fixas pristinas] and to slope towards other succeeding stars, as can be judged in general. For that movement is so very slow that within the 1400 years from Ptolemy to us this cannot be safely enough affirmed of all the planets.

Perhaps those axes of rotation of the bodies play the roles of the threads which you bring in here instead of librations

The axis of the daily rotation of the earth—of which I have spoken in the doctrine of spheres—forever points in longitude towards the beginnings of the Crab and of the Goat. For this axis prolonged in both directions marks out the poles of the world, as in Book II, page 150. But the arc drawn from the pole of the world perpendicular to the ecliptic passes also through the poles of the ecliptic: therefore it is the colure of the solstices and marks out the beginnings of the said signs.

But the threads by which the Earth is repelled from the sun or attracted pass from sign to sign. The aphelion [586] of the Earth was formerly in the Archer, but now it is at 6° of the Goat. Therefore the axis of rotation of the Earth and the thread which changes the interval are different.

Then it seems that the Earth nevertheless should be at its greatest distance in the beginning of the Goat. For if the whole body of the Earth rotates around that axis, the thread will be rotated too, in so far as it differs in posture from the axis, and it will describe as it were two cones with their vertices meeting at the centre of the Earth, and only at one moment of the day will it look towards its proper place: during the rest of the day it will revolve around the beginning of the Goat pointed out by the axis of the Earth. And thus it will pile up all its own force at this axis, and by a certain as it were spiral line will draw the Earth away from the sun and always towards the region pointed at by the axis.

Certainly in this way, by the tight connection of the thread with the axis of daily movement, what is spoken of would take place, and the apsis of the Earth would never depart from the beginning of the Goat. Or else we are therefore compelled to admit that the globe is inside an outer crust: in such fashion that the crust rotates during the daily movement, while the globe having the threads does not rotate; and the ordinary magnetic virtue belongs to the outward crust, because it always shows the poles of daily rotation but not the apsis of the sun or Earth.

Whence let some physicist come to the help of J. C. Scaliger, who argues about the rising of rivers and the ebb and flow of the sea; and let the physicist see if these separated bowels of the Earth can aid him in his labour. Even though the moon and the soul of the Earth are sufficient for me.

If the planetary globes have an inward rectilinear magnetic configuration, why do you not rather ascribe to them themselves the reason for their fleeing from the sun and approaching the sun, in proportion to the diversity [587] of the regions of their body, as was done in the Commentaries on Mars?

1. Because astronomy bears witness that this drawing away from the sun and drawing near to it takes place in a line so to speak extended towards the sun, in so far as the intermingled revolution does not vary the line. But the magnetic threads are rarely stretched out towards the sun.

2. Because two very diverse things would be attributed to those magnetic threads. For first, they would point towards the same region of the world—which is something like rest; then they would move their body in place, now away from the sun, now towards the sun. But by means of repulsion and attraction this [movement of approach and withdrawal] is united more simply with the laying hold and making bodies revolve, which the sun furnishes.

3. Furthermore, it is more probable that the form of the solar body and its virtue continue as far as to the planets than that their form and virtue continue as far as to the sun, so that, when repelling it, they flee, and when attracting it, they seek. For the sun is a huge body; while a planetary body is quite small. The light and heat of the sun manifestly descend to us; the sun makes the planets revolve. Accordingly before this, things were clear concerning other virtues of the sun. But we do not have such and so evident testimony concerning the prolongation of planetary virtue as far as to the sun.

4. It will be shown below that the threads of the body suffer some slight deflection on account of the sun. Therefore it is probable that the libration of the whole body also comes to it from the sun rather than that it is inborn; that is, it is a passion from another, not an action or movement from the planet itself.

But would you set down that this virtue is at least shared between the sun and the planets and that the force of repulsion and attraction passes back and forth from one to the other, just as it is shared between two magnets?

No: rather this is the fifth reason why this repulsion and attraction is not attributed to the planets themselves: in order that there be no mutual attraction and repulsion according to [588] the very institute of the Creator, who does nothing in vain. Therefore, if the virtue of the planet extended as far as to the sun, then the planets, in the direct ratio of the bodies, would move the sun out of the position which it occupies at the centre of the world, or at least the sun ought to stagger, attracted now in this direction, now in that, according as many planets having a like faculty assail the sun from one side.

You seem not to avoid the following inconveniences: for the sun, leaning against the form and virtue of its body as against a pole and pushing the planets, pushes itself out proportionally, and drawing the planet as it were with its claws, it similarly draws itself to the planet?

We have avoided this in every way by denying mutual attraction and repulsion. For firstly, neither the shape [*forma*] nor lay out [*dispositio*] of the bodies will be directed towards this, if such a virtue of the planet [589] does not extend to the sun. Then something such does not actually follow—as though short of the Creator's design—from material necessity alone. For the bulk is so great, the density of matter in the solar body is so great, and its force of attraction and repulsion is so great; and in turn the weakness of the planet and the weakness of its resistance are so great; that the sun is in no danger of losing its position. Thus when a ship sticks fast in the sands and cannot be torn away and moved from its place except by two hundred horses, nevertheless one hundred horses, though they are half of the required virtue, do not move forward half of the lone thing, because there is no mean half-way between the moved and the not-moved, since they are contradictory.

State a convincing hypothesis as to how any planet completes its circuits and is meanwhile attracted and repelled.

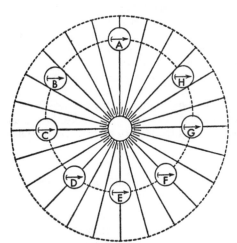

Let us begin with that moment when the magnetic threads offer their sides to the sun, so that both extremeties of the threads are equally distant from the sun; and this takes place, by the foregoing diagram, at *A* the greatest distance of all: at this time the sun is neither repelling the planet nor attracting it; but, as if hesitating between both, it nevertheless lays hold of the planet; and by means of the rotation of its body and the flowing out of its form, the sun makes to move forward from *A* to *B* the planet which it has laid hold of, and the sun overcomes its resistance and in turn is overcome by the planet, so that the sun lets it fall, so to speak out of its hands, *i.e.*, out of the preceding rays *A* of its form and virtue, and takes it up again with the following rays *H*, and does so in the fixed ratio of the virtue of the form in that interval. In this way the planet is moved forward, while in the meantime the magnetic threads, by the force of direction, look towards the same region of the world, in such fashion that the region friendly to the sun is gradually turned towards the sun, and the discordant bends away from the sun: at that time therefore the globe begins to be attracted—only a little, if there is little difference between the distances of the extremities from the sun: by this attraction the planet will go, from the wide circle begun at *A*, gradually inward to *B* and will betake itself towards the sun, as if into a narrower circle and a stronger, because denser, virtue [590] of laying hold: hence it frees itself less easily from the virtue and so is moved on more speedily. This attraction, which is very slow at the start very near to *A*, is very rapid at the time when the sun has the whole friendly hemisphere of the planet within its view, but the whole discordant hemisphere is hidden behind the planetary body, *i.e.*, when the magnetic threads point in a straight line towards the sun; and that takes place around *C*, a quarter of the whole circular ambit. Thence around *D* once more this attraction towards the sun becomes more lax; but the velocity of the progress along the circle continues to increase, because the interval between the planet and the sun is still decreasing on account of the attraction. This relaxing of the attraction amounts to almost nothing in the beginning after *C*; soon, it is felt more and more, the more the unfriendly part of the planet thrusts itself out and offers itself to the view of the sun, towards *D*, until half of the circuit has been traversed and once more both hemispheres of the revolving globe regard the sun equally. For at that time all attraction stops, and the planet is very near to the sun, and so is very speedy, because it is struggling with a very dense and a very strong virtue of laying hold and frees itself hardly at all from that encompassing virtue.

But immediately the globe is borne beyond this place *E* on its orbit towards *E*, because already the discordant hemisphere is nearer the sun than the friendly hemisphere and is advancing more and more towards the sun. The planet too begins to be pushed away from the sun, as though out of a narrower and denser sphere of solar form into a wider, more rarefied, and weaker sphere. Whence the decreases in its movement follow, and in the contrary order: first, slowly, after *E* towards *F;* then, where the whole discordant hemisphere or region of threads points in a straight line towards the sun, but the unfriendly region is turned away from the sun: the planet is repelled very swiftly, but its movement has now once more slackened to middling. This again takes place around *G* the other quarter of the circuit. When the planet has been borne farther towards *H*, this repulsion again slackens, until it completely disappears at *A*, where the planet has returned to its primal place and has been repelled to its greatest distance from the sun.

[591] *But it is unbelievable that the planet—this freedom allowed—should, after completing its return, be restored to exactly the same distance.*

Doubtless this is at last a good place for that pronouncement of Ptolemy's written out above, where he reminds us that nothing takes place in the heavens which hinders the natural movements of any body or which makes the bodies stray as it were from their foot-paths. And so, if such laws of movement have been instituted by nature that the planet return unto itself most exactly, then this will most certainly take place, although without the shackles of spheres, in the free ether. But the laws were made as we described them. For the halves of the circuit are equal to one another—the one in which the planet is attracted and the other in which it is repelled. Equal times are taken by both halves. Moreover, the virtue of the sun is the same and everlasting—both as attracting and as repulsing. And it has the same ratio to the planetary inertia, which is always the same, because in an everlasting body. Accordingly it does as much by attraction in one half, as it does by repulsion in the other. Why then should we be doubtful concerning the restitution of the planetary body to the original distance within one period of time?

Also in these terrestrial and violent movements, are not the movable bodies separated from that which was the cause of movement, as in scorpions, ballistae, catapults, bombardae, and slings? And weapons hurled fly through the free air, and nevertheless they reach their appointed place. And, for a wonder, there are some sclopetarii and slingers with a sureness of aim that cannot be imitated. If in this case the form of that movement—which movement was in the thrower at the movement of throw and was directed towards a fixed region—the form impressed into the movable body for a short time and vanishing has such power that, as long as the movable is carried by a form which has not yet become completely feeble, it does not cease to strive for its appointed region: by how much firmer defences will the certitude of the celestial returns be protected, which are governed [592] by the inward and plainly united and hence everlasting threads of the movable thing—since in the first case the air disturbs the movement by its impact and encounter; and in the second case the density of the ether to be passed through is clearly nil in effect or else has a very slight effect.

Why are not the librations of the different planets in the same ratio to their mean distances, that is, why is the eccentricity of Mercury greatest, next, that of Mars, and then those of Saturn, Jupiter, and the Earth, while that of Venus is least?

The instrumental cause is the different strength of the threads, whether that is produced by nature or by posture. But the final cause is the same as that of the eccentricities themselves, namely in order that by reason of these eccentricities the movements of the planets should become very fast and very slow in such measure as would suffice for the harmonies to be exhibited through them. Book v of my *Harmonies* has to do with this.

There remains one difficulty with respect to the direction of the threads towards the same region of the world. For since you said that one region of the threads has friendship with the sun, and the other is in discord with the sun, so that in conformity with the former or the latter region the sun either attracts or repels the body of the planet, it seems that the sun has the power over the planet to do what is less, namely to move these threads out of their parallel posture and to turn them towards itself, before the planet is transported into a position from which the threads can look towards the sun.

There is nothing absurd in something like this taking place, so that the sun struggles with the direction of the threads, just as it struggles with the inertia [593] of the body with respect to movement in place—provided we keep in mind that the sun accomplishes less with respect to deflecting the threads than in moving the whole body in place, just as it also accomplishes less in respect to attracting the planet. And this tempering pertains to the Creator's plan, lest the planets should come into contact with the sun, if they were not transported into the opposite half of the circuit in a shorter time than the whole interval of time which could be spent in the direct attraction of the thread.

Therefore, since the circling of the planet around the sun forestalls the deflection of the threads, hence, although in one quadrant of the circuit the threads deflect somewhat towards the sun in their friendly region, and away from the sun in their discordant region; nevertheless, because the planet is transported into the other quadrant before the deflection of the threads becomes total, it follows equally that there is a change in the posture of the contrary regions, which are turned towards the sun, just as if the threads were not deflected; therefore in the remaining quadrant the sun by the same force gives a counter-deflection in the other direction to the planetary threads, which lie in the contrary position and turn their unfriendly region towards the sun. And so by the earlier contrary deflection the sun again restores the planetary threads to their parallel posture. In Book v this deflection and counter-deflection become the principal means of calculation.

Could you cite a common example of this direction and mixed sloping of the threads?

There is an example in the magnetic needle. For although it looks towards the north if it is free, nevertheless it is somewhat deflected from the north if a magnet approaches obliquely: for then it bows somewhat towards the magnet.

What things are required for the perfect restitution of the threads to the parallel posture?

That the sun expends as much of its forces in deflecting—say, through the quadrant PIN, attracting H the solipetal region of thread downward from line IS towards itself— as it expends in restoring, as through the quadrant NER, drawing back G the same region of thread upwards towards line SY, which is nearer to itself. But this can take place only if, with PR as the line of the apsides and PN and NR as full quarters of the orbit, NQ the thread of the planet, situated at N, the limit of the quadrants, points precisely towards the sun A. For even though, in the upper quadrant PN, the sun A administers this deflection SIH, BNQ, at long distances AP, AI, etc., and so with a feebler virtue, but in the lower quadrant NR at the short distances AS, AR and so with a stronger virtue; yet in turn the planet delays longer in the upper quadrant PN and undergoes those feeble forces of deflection for a longer time; while in the lower quadrant NR the planet makes a shorter delay and has a shorter time in which to undergo the strong

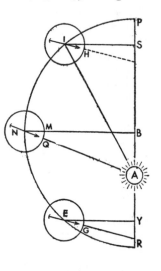

forces which counteract the deflection. And there is perfect counterbalancing. For the same perfect counterbalancing can also cause that at N, one and the same limit of the quadrants, the distance AN—in the rightly drawn orbit— should be equal to the semidiameter BP—as will be made clear in Book v.

But what if the planet were to direct the thread NQ towards the sun not precisely after traversing PN the upper quadrant of the orbit but at some later time?

Here the burden of proof is on the opposition. For where the thread points towards the sun is the limit of the quadrants, which are measured from the apsides. For IH, the deflection of the thread with respect to the perpendicular IS, is always increasing, as long as the thread H seeks the sun. But the increment of libration increases with it—the effect with the cause.

[595] Therefore if, in this work of attraction of the planet towards the sun, more than a quadrant of the orbit is traversed with reference to the fixed stars; then more than a quadrant will have to be traversed by the planet in re-

storing the right angle between the thread and the sun at R and in the effect [of the restoration], or the remaining part of the libration, whereby the planet is brought from the nearness NA to the nearness RA, by the same degrees of increments which are now decreasing in the contrary order.

Therefore, when the quadrants are added together, their excess over the semicircle shows the magnitude of the change in direction of the threads under the fixed stars in half of one period, or the magnitude of the transportation eastward of B the centre of the orbit and PR the line of the apsides. Therefore if this magnitude is subtracted from that which is more than half of the orbit as estimated with respect to the fixed stars, there will remain not more than half of the elliptic orbit as reckoned from the apsis P.

Then do the apsides abide, or are they transported from one place to another under the fixed stars?

In the case of Jupiter the comparison of observations made by the ancients with those of today bears witness that the apsides stand approximately under the same fixed stars, or retrograde very little. But in the case of all the remaining planets, the apsides are found to have left their original seats, by a movement in the direction eastward—as in the case of the apogee of the moon—but in the case of the planets by extremely slow movements, though the apogee of the moon progresses quite perceptibly.

What is the reason why in the case of the primary planets, the threads are found so perfectly restored—after the whole periodic returns have been completed—that the progress of the apsides is imperceptible?

Because it is the same sun which gives a libration to the planetary body and

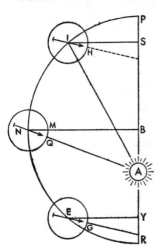

which deflects and restores its threads; and because in both acts it is the same threads by which as by instruments the planet is given a libration and is deflected: then there is no cause why the forces for both acts should not be measured out through equal times. For just as the planet, which points its thread NQ directly [596] at the sun, would at some fixed time be joined even in contact with the sun, if it did not leave the line NA; so too the same planet, placed at the same N and, by a fiction, directing its thread at right angles to the line NA, would be turned around together with its thread in a fully equal time, so that in the end it would direct the thread NQ towards the sun. But just as to the libration there is added a third work, namely the moving of the planet out of its posture AN, so that the thread NQ no longer points towards the sun and hence is not attracted with the same strength towards the sun—and in this way it is seen too that full contact does not take place by a movement along NA, but is forestalled by the transportation from N to R, and the planet gets no nearer than RA; so too this same transportation of the planet from N to R anticipates this deflection of the thread, so that the thread will point at the sun long before it could have been rotated a whole quadrant by the sun; and so, instead of the quadrant of rotation, no more than arc QM is needed. But the angles of the past deflections, or the virtue expended in them, are measured by the sines—as will be made clear by examples of natural things in Book v. Wherefore as PB the total mean distance—or in the ellipse, NA—is to BA the magnitude of half the libration—the work of one quadrant; and this magnitude is the same as the eccentricity—so also NQ the semidiameter of the planetary globe, which is employed as the whole sine, will be to the sine of MNQ the angle of greatest deflection; and MNQ will be the angle of greatest deflection at the time when, by the transportation of the planet, a quadrant from P the place of greatest distance PA has been exactly completed.

But with the assumption of this proportion, it is demonstrated that the thread NQ points towards the sun A at the time when PN is a quadrant under the fixed stars, precisely. For let AN be equal to PB, [597] as in the ellipse; and let B be the centre of the eccentric circle, and ABN a right angle, because its measure NR is a quadrant. Now from Q the solipetal limit drop the perpendicular QM upon BN. Two right triangles ABN and QMN are formed. And because it is assumed that $NQ : QM :: NA : AB$, then N, Q, and A will be in one straight line, or Q will point towards the sun.

But it has already been demonstrated above that if, when PN a quadrant under the fixed stars has been completed, the thread Q of the planet points towards the sun, so that BNQ is its angle of deflection; then it follows that in NR the other quadrant under the fixed stars the thread NQ is restored and BNQ the angle of deflection is annulled. Hence when the planet stands at R, the thread is once more parallel to BN, as it was at P. And this restitution of threads is complete after the semicircle has been traversed. Let the same judgment apply to the other semicircle; for when that has been traversed, the planet returns to the same position under the fixed stars.

But since experience bears witness that the apsides are transported imperceptibly, and do not remain under the same positions among the fixed stars; then it follows that NQ *looks towards the sun at not precisely a quadrant from* P *the original position of the apsis. What is the cause of this straying from the already established ratio of regularity?*

It seems that the imperceptible slowness of these movements must be sought for in material necessity—if anything else is—namely in the straying of the said movements, the libration and deflection, from one another through the intervention of a third movement. For it diffuses itself in an infinity of time, which has no beauty, as being indeterminate. But it is difficult to state what the intervening cause is: because all astronomers are not decided about the thing itself; nor, for most of them, does the thing have any fixed magnitude. But with magnitude removed, we lack any means of examining the causes (which some one might have searched into by conjecture) of whatever sort of latitudinal digression the planets may have away from the ecliptic. For the digression does not take place without the deflection of the threads NQ to AN the ray of the sun—a deflection as great as the digression of each planet. [598] It is consonant with such a greater or lesser deflection that the work of the threads should be somewhat weakened—and that variously, in proportion to the varying relation of the digressions to the apsides. In Saturn, Mars, Venus, and Mercury the mean longitudes[1] have some latitude, but none in Jupiter; and the apsides of Saturn, Mars, Venus, and Mercury progress in that proportion, while those of Jupiter stand still. Therefore since in other respects the force to deflect the thread of the planetary body becomes greatest at the apsides P and R, where the thread is presented to the sun at right angles, it is believable that this force becomes slightly weaker on account of the latitude. And the reason the force does not suffer the same loss from the libration too is that there the libration is almost nil in itself. In turn, the force of deflection at N is almost nil, while the libration is greatest: therefore the force of deflection suffers loss from libration here but not there, in proportion to the latitude. And it can happen that the deflection of the thread can be slowed up in this way; and with that given, what has already been unfolded takes place; the thread looks towards the sun more slowly, namely beyond the boundaries of the quadrant. But it has been demonstrated before that at that time the apsides are transported eastwards. Then, this can be the cause of the said phenomenon—a cause linked to physical or geometrical necessities, in accordance with the principles previously laid down.

2. But in the meantime I would not rigidly deny that this effect can be a part of the design, so that it is not a consequence of necessity, or a mere consequence: because we are still ignorant of the magnitude of it. Then there will be room to speak about the final cause: to the final cause belongs the mutual tempering of the forces of libration, of the deflection of the threads, and of the revolution, in some fixed proportion: in order that, because the librations were prepared in order to set up the harmonies of the movements, any given harmony should not be born always in some one configuration of two planets, but in the succession of the ages would pass through absolutely all the configurations, and in order that thus all the harmonies of movements—which Book v of the *Harmonies* is about— should be mingled with all the harmonic configurations—which is the matter of Book iv of the *Harmonies*.

NOTE BY KEPLER: Pages 102-104. A new and hasty correction has perverted the original and well-meditated text: 1. There is a *petitio principi*. 2. It is not in my design that on page 102, fourth line from bottom, there should have been a case of necessity: the direct variation of the libration with the deflection. 3. In the same place, the cause is not a cause. 4. One thing was proposed on page 102, and something else was demonstrated on page 104: on page 102 it was a question of the fixed stars: and on page 104, of the apsides.

The true cause of the almost perfect restitution is of physical necessity. For either the threads remain parallel, or else in one half they are deflected downward away from the apsis, as NQ, and in the other half, upward, since in both directions the counterbalancing is perfect, as was said on page 102; moreover thus the threads are parallel to one another at both apsides; therefore the restitution is perfect.

Therefore on page 104 something false and contradictory is proposed: the straying of the libration from the deflection. Instead, the cause was to be given which is suggested on pages 101-102. For in the upper quadrant PN the sun gives a slightly smaller deflection, in the lower quadrant NR, a slightly greater counter-deflection; if at any rate the fixed stars have their termini at the quadrants. Accordingly since the solipetal terminus G is at R, the point of the fixed stars, and is already above SY, and is therefore still nearer the sun; therefore the planet is still sailing on: wherefore R the perigeal apsis will be beyond R in the fixed stars. If the latitude of the planet is

---

[1] For *apsides* reading *longitudines mediae*. See note by Kepler below.

the cause of this thing, it will have to be explained in another way than on page 945, where instead of apsides read mean longitudes, because in Jupiter P is not the node but the limit. Nor is it sufficient to look to Jupiter and to say the reason the apsis is standing still is that the apsis is at the limit. But it is necessary to explain this too: why the progress of the apsides in the case of the other planets of very unequal periods is approximately equal under the fixed stars.

4. ON THE MOVEMENT IN LATITUDE

[599] *Under what laws do the planets digress in latitude away from the ecliptic?*

Again, under a very simple law: that the plane which they circumscribe with the centre of their body be exactly even in any period and that it be inclined to the plane of the ecliptic in a constant and invariable inclination—except in the case of the moon.

If level planes are inclined to one another, they meet and cut one another in one straight line. I ask what that common line is, from which the orbit of the planet is inclined to the plane of the ecliptic.

In all the planets, this line passes through the centre of the sun; and the line of each planet extends to its own proper places on the ecliptic, which are opposite to one another from the centre of the sun.

How is this established?

Because, since the planet at two different points on its return, as at C and D, is seen to be under the ecliptic without any latitude; these two positions on the orbit are found by calculation to be in the same straight line CAD with the sun A; so that if ACM was at 17° of the Bull, the interval of time until the planet was seen again on ecliptic, together with the hypothesis of the eccentric circle, exhibits the line ADO of the other position on the eccentric circle at 17° of the Scorpion, that is, opposite 17° of the Bull.

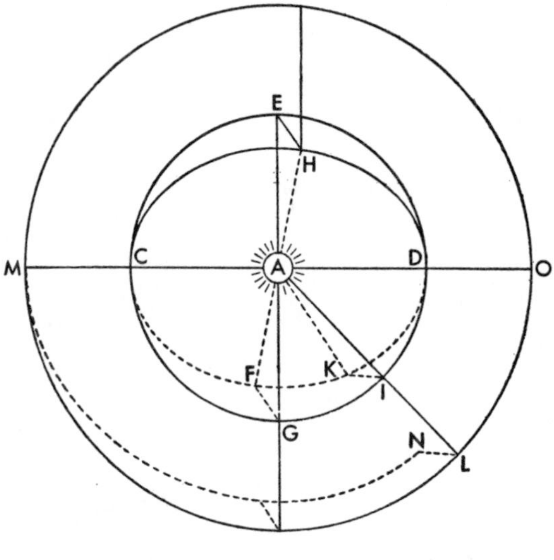

What is gathered from this?

The same as above, on page 70. For since the planes of all six eccentric circles meet at the one common centre of the sun; then all cannot meet at once anywhere except at this centre of the sun, because the line of section is not common to all, but [600] proper to each planet. But different lines cannot meet at more than one point.

Accordingly because the sun is the node common to all the systems: therefore either nature moves the planets by bodily virtues, or else mind, by rational commands. For the planets the sun is a fixed mark, which all their revolutions regard.

What causes do you assign for the movement in latitude?

In the case of the planets the sun is not the cause—unless the remote cause—for this deviation from the plane of the ecliptic; nor is there any need for the planet's intelligence in this work, nor for the above disproved substructure of solid spheres, upon which as upon chariots the planets travel along their orbit; and there is much less need of the spheres here than in the case of the librations into opposite altitudes or in the case of the movement in longitude. But some certain formation of the planetary bodies is alone sufficient to twist their orbits away from the ecliptic and to twist them back again.

Why is not the sun ranked as a cause, since it has already been said that the lines of section go through the solar body itself?

[601] Because one and the same sun, by means of one and the same form of its body, which form revolves in a uniform and very direct current along the mean circle between the poles of rotation of the sun, cannot carry different planets through other different paths, unless the planets add from out of themselves the causes of this differing digression in latitude.

Of what sort do you hint that this formation of the planetary bodies is?

It can be either essential, that is, belonging to the inward rectilinear magnetic thread, or else accidental, namely, the rotation of the planetary globe around its axis performed in such a way that the threads or axis of rotation retain their parallel posture during the whole circuit of the body and have such a direction that, when the planet is on the ecliptic, it touches the orbit and deflects at one terminus somewhat towards the north, and at the other towards the south.

Do you have a common example of this deflection?

The oars of ships supply some sort of example. For if the ship is being driven forward by winds, but an oar is tied obliquely to the stern, then the ship against which the line of wind is bearing is turned gradually to the side.

The oar, pole, or rudder always directs the ship towards one single region. How therefore do the planets now depart to the sides of the ecliptic and now return from there to the ecliptic?

If the rudder of the ship is turned, the ship too deflects to the other side. Even though the planets keep their threads straight, in a parallel posture, and unturned, nevertheless the planets are transported to the opposite parts of their circuit, wherein the threads, [602] by reason of their original posture, have the opposite inclination to their orbit; wherefore too the planets in the other semicircle are driven towards the opposite regions.

In order that I may better understand this movement, state what surface one such thread or axis produces during the revolution of the planet around the sun?

Let us lay down that when the planet is on the ecliptic, as here at C and E, then *AB* the thread of latitude does not have any inclination with respect to the sun—though this can take place differently but with the same effect, if the posture is equivalent—but has such an inclination with respect to the plane of the ecliptic, that *EA* or *CA* the half [of the thread] is to be understood to be sunken below the paper—which represents the plane of the ecliptic—but the remaining half *EB* or *CB* stands out above the paper, and the angle of inclination is as great as the latitude is accustomed to be at the limits—*F* the limit above the paper and *D* the limit below. Moreover, let the movement of the solar form, as if of a river or wind, be from *E* towards *F*, *C*, and *D*.

[603] Then since this movement at *E* is going to advance contrary to the sunken half of the thread *AE*, but at *C* is similarly going to advance contrary to *BC* the half which stands out and which is opposite *AE*. Furthermore, at *E* the movement pushes the planet up above the paper, in the direction whither *B* the front terminus tends; but at *C* it pushes the planet downward below the paper in the direction whither *A* the terminus in front of that position tends. In a rudder the opposite takes place, because the rudder is pushed by the force of the river, not driven by an inborn aptitude. But since meanwhile the thread *AB* remains in a posture parallel to itself throughout its whole circuit;

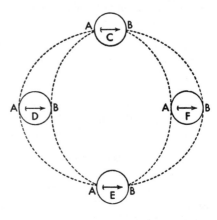

hence, when the planet is most northern at *F* or most sunken in the south at *D*; neither terminus *A* nor terminus *B* is in front, but the thread *AB*, which stretches out as it were into the depth of this river, *i.e.*, towards the sun, and which receives the attack on the right side, furnishes no cause for further removal into any region: wherefore a changing around takes place at these points, so that, although before point *F* the terminus *B* had been in front, now after point *F* the terminus *A* is in front; and so the planet begins to approach the ecliptic once more, at first with imperceptible progress.

From this it is now clear what sort of figure is engendered. For because the thread *AB* is moved from *E* towards that very region towards which the front terminus *B* tends; then the surface which is produced by *AB* is at point *E* diminished to a mere line, which nevertheless gradually becomes a surface, and, having arisen from the point *E*, acquires at *F* its greatest latitude, equal to the length of the thread *AB*: thence once more the surface is diminished as far as *C*, the segment of the circuit which is opposite *E*, the first-mentioned segment: there again the surface disappears into a line. The same things are to be understood of the opposite semicircle *CDE*. But the thread borne thus to *F* and *D* at an inclination and always following its own lead will produce a perfect plane, *i.e.*, in so far as it keeps its posture parallel; and if this plane is continued, it will pass through the centre of the sun because the thread *AB* looks towards the sun—at *F* with the terminus *A*; but at *D*, with the terminus *B*.

But with this continuation of the plane removed, if that which is produced by the thread [604] is alone considered: it will be such a form as two little crescents exhibit between the two ellipses, *BCAE* the outer ellipse, and *EACB* the inner, which touch one another at *C* and *E*, so that the same line *CE* is the major diameter of the smaller ellipse *EACB*, but the transverse diameter of the greater ellipse *CBAE*.

Moreover, the centre of the planetary body will revolve in a perfect plane, which in this figure was made circular, namely *CDEF*, although the plane itself also, as is clear from what was said above, bends away from the .perfection of a circle to an ellipse with its lopped-off sides.

The oar or rudder of a ship stretches out from the ship straight into the waves or wind: but these threads lie concealed within the round body of the planet: therefore it is not the same force which is in rudders.

It is not necessary that all things should correspond in an analogy. But in place of the oars there is the other much more suitable force in the threads. And just as, above, the threads have a natural inertia contrary to the deflection of themselves, or rather a power to keep their posture parallel during the transportation of the body; so now also there is present in the threads of latitude, in addition to the similar force to keep their posture parallel, a natural power of agility too, or of following exactly the same line and of directing along that line the movement produced in the planet, in so far as the movement tends towards the same region as the other extremity of the thread.

Compare this form of movement in latitude with the ancient astronomy in an everyday example.

Here we entrust the planet to the river, with an oblique rudder, by means of which the planet, while floating down, may cross from one bank to the opposite. But the ancient astronomy built a solid bridge—the solid spheres— above this river—the width of the zodiac—and transports the lifeless planet [605] along the bridge as if in a chariot. But if the whole contrivance is examined carefully, it appears that this bridge has no props by which it is supported, nor does it rest upon the Earth, which they believed to be the foundation of the heavens.

Nevertheless is not this theory of the movement in latitude more difficult than if some- one imagines solid spheres?

But, reader, you ought to remember that we are here busy with the physical theory of causes, on account of which any hypothesis is applied in order that we may know what of truth exists in such an hypothesis or astronomical fiction. But below, in Books v and vi, for the sake of understanding, we shall not reject the whole circles and their inclinations to the ecliptic; because they are equivalent to these attractions of threads to the sides of the ecliptic.

If that earlier libration of the planet in altitude and this digression in latitude had the same boundary-posts on the ecliptic and were effected by the same threads, the causes which you assign would be probable.

Indeed, what prevents one and the same globe from having twofold rectilinear threads stretching out over the whole body, so that from some it receives a libration in altitude, and by the others is rowed backward and forward? Thus

on the surface of rivers a threefold movement of the parts is
discerned—each movement proceeding in its own direction:
the first is the flow of water, the second that of the waves,
which that flow casts crosswise to the banks in a continuous
series, the third comes from the wind. For if a contrary wind
blows obliquely, it roughens the surface of the waters, and
starts another series of smaller waves moving in their own
direction, and these waves advance on top of the earlier waves, which are not
troubled. Thus above in Book I [606] I mentioned the nature [*substantia*] of the
belly, which represents a triple-woven tunic and contains three kinds of threads,
distinct as to their regions, the seats of the three faculties of attraction, retention, and expulsion: although the weaving belongs to not one but three tunics.

*Do not the farthest digressions of the planets always occur at the same positions on
the zodiac, or do even these digressions change their positions?*

Observation of the progression of the limits is more difficult still than that of
the apsides: nevertheless they are seen to be going back gradually westward in
the sphere of the fixed stars and more slowly than the apsides are progressing;
there is a case of both in the movements of the moon.

*If the limits retrograde and the apsides progress, will the threads productive of
latitude remain inwoven with the threads of longitude, both of which you have given
to the same globe?*

This retrogradation clears our way to the inward substance of the globes, into
which narrows we have already been compelled to go, in the comparison of the
daily revolution of the Earth with its threads of libration. Accordingly here too
we can seek within the outer crust a separate globe—like the yolk within the
white of egg—supplied with its own threads, and able to revolve according to
the same laws, and with the strength of its forces different from that of the outer
crust, if need be; so that both of them [the crust and the globe] can be deflected
by the same outside cause, with different measures of speed, if there is need of
this.

For thus in the case of the belly which I have already brought in, there are
three tunics: the outmost, the inmost, and the middle; and one of them can be
passive, if the others are injured, or be active, while the others are at rest; although they are unlike this thing, in that they are not separated from one
another.

The ancient astronomy places solid and plainly adamantine spheres on top of
other spheres, where no body is visible to us and the whole region is as transparent as if it were empty. [607] Therefore it will not take offence if we construct
something similar in the globes, which are visible and palpable bodies.

*Cannot this already mentioned axis of revolution of the outer crust of the planetary
bodies perform this office of deflecting the movement of the planets latitudinally?*

This cause rests upon very great probability, as will have to be said in Books
VI and VII in the explanation of the schemata of the sun and the eighth sphere.
Nevertheless nothing certain can be affirmed of all [the planets]: because even
though we said it was believable that the remaining primary planets rotate
around the axes of their bodies, nevertheless the regions towards which these

axes tend or incline are unknown to us. Wherefore we have an example in the Earth alone. And the moon, a secondary planet does not rotate, although none the less it completes its latitudes.

How can it be effected that the limits of digressions retrograde westward?

In Book VII part of this appearance will be explained as accidental, not as physical or real. But what remains of this movement and is real is caused by the deflection [*nutu*] westward of the threads of latitude: so that they remain in one and the same plane precisely, during their whole circuit, but the threads themselves—*i.e.*, the globe itself—are deflected backward over the centre of their body secretly in accordance with these threads.

From what causes does this counter-deflection arise?

Up to now the probability of most of the causes brought forward was clear. But in this last train of astronomical matters, the causes have ill success; and both understanding and belief in those things which one is able to devise have to work hard. Let us however speak of as much as we can find out. [608] We have said that the nature of the threads of latitude consists in an aptitude to move forward into the region of their own parallel direction. We have said moreover that while the planet is being transported from C or E, the position which it has on the ecliptic, into D or F, the position of farthest digression north or south, in the meantime those threads remain parallel, and for that reason it comes about that, since there at C and E the threads touched the orbit, now here at D and F they are sunken in the depth towards the sun, whither that movement to which they are deflected does not tend; rather at that time, the motor river from the sun, so to speak, meets the crosswise threads AB at right angles, more speedily below—that is, at A in the position F and at B in the position D—than above and on the outside. Therefore if the threads are deflected towards the movement, it is surprising if this deflection, which with its lower side seeks the region of the movement, takes something away from the paralleleity and does so at both limits? So the retrogradation of the limits follows, since no counterbalancing exists. For A is pushed out, at F, along the road EAC; and B is pushed out, at D, along the same road CBE: so in both cases B will be deflected below the paper.

But if this cause is not admitted, then let the motor soul be summoned to twist together, by its laws, an inside kernel within the outer crust, according to the design of the Workman that by the interweaving of the orbits with one another and their frequent plurification and condensation the solidity of the sphere might be somewhat penetrated by the planet in the succession of ages.

Why is the regression of the limits slower than the progression of the apsides?

Although there is doubt concerning Mercury, there is also some doubt concerning Jupiter: nevertheless let us follow probability on account of the clear example of the moon, and let us say that the cause is as follows: because a disturbance, if there is any, coming from one and the same outside cause is necessarily more perceptible in a great movement than in a small. Now the transposition of the apsides arises from a great movement, which is the deflection and counter-deflection of the threads in any semicircle, [609] and is as great as the optical additosubtractive difference [*aequatio*], and it would become greater and altogether total if it were not anticipated by the revolution of the planetary globe. But the transposition of limits takes place through a small movement, the digression of a few degrees in latitude, and is not greater than this measure of it, so that this digression can give rise to nothing else whereby the transposition of the limits may be hindered. Wherefore the same rays of the sun which introduce movements in both cases, by the laws already explained, have more manifest effects there [in the transposition of the apsides] than here [in the transposition of the limits]. In addition there is the fact that in the first case rays of the sun act in relation to a greater diversity than in the second, other things being equal. For in the first case the obliquity of the rays of the sun with respect to the threads—and this obliquity tends to the side—or the angle of latitude, whereby the work [of the threads] is weakened, was perceptible. In this second case the diversity between the parts of the planetary globe, and hence between the termini of the threads of latitude—the terminus nearest the sun and that farthest away from the sun—to which diversity we assign the movement of the limits, is mighty small. Accordingly by right this second work is less than that.

5. On the Twofold Irregularities of the Moon and their Causes

These things which up to now have been argued concerning the causes by which the true movements of the primary planets are made irregular are not also to be understood concerning the moon, a secondary planet, are they?

1. On the whole, the moon imitates the same general form of movement, around the Earth, as that which the planets keep to around the sun; and so too we ought to set down the same causes within the lunar body, that is, threads of magnetism, their rectilinear tract, and the contrary regions of that tract—one region friendly to the Earth and the opposite region unfriendly—and finally the approximate parallel-eity to itself of this tract throughout the whole circuit of the moon. [610] Accordingly, when the moon has been transported into the opposite position, there is a changing around of the regions; and in

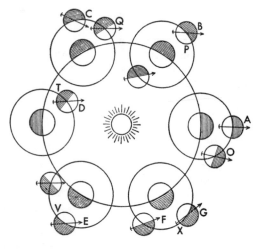

conformity with the friendly region the moon is attracted by the form of the terrestrial body; and in conformity with the unfriendly region it is repelled. And its movement in longitude too is increased or decreased in that proportion; and similarly we must conceive of other threads in its body, by means of which threads the moon's digressions from the ecliptic are produced.

In this diagram are represented some positions of the Earth going around the sun, together with the surrounding lunar heaven; and in the heaven of the moon some positions of the moon going around the Earth. But the image of the magnetized needle signifies the magnetic threads in the lunar globe, by reason of which the moon becomes eccentric to the Earth. For the points A, B, C, D, E, F, and G signify the region friendly to the Earth, and they lie at the second mean longitude, and so at A and G the moon is situated at the mean position between the perigee and the apogee; [611] at B, Q, and D it is a little before the apogee; at C, a little after the apogee; and at E and F, a little before the perigee.

2. But since this movement of the Earth, as was explained above, is derived from two sources, as it were—namely, from the daily rotation of the Earth, and from the rotation of the sun around its own axis—and this sun is the midpart of the great sphere common to the Earth and the lunar heaven; so it will be reasonable that the true and real movement of the moon around the Earth—and in so far as that revolution around the sun, which is shared by both and is accidental to the whole lunar heaven, is removed from it mentally—has two sources, as it were, and suffers all those affects as twofold which the movements of the primary planets have as simple. And this is perfectly consonant with the experience and skill of artists, and with the words arising from this twofoldness. For not merely in the [pages] above, according as there was a solitary mean movement in some one of planets, was there a mean movement and a semimonthly variation of this mean movement in the case of the moon; but also here [hoc loco], where it is a question of the periodic irregularity of this movement, and this irregularity is not semimonthly like the variation but monthly or rather semiannual: we find in the moon a twofold intensification and remission of mean movement at the contrary moments of the period, instead of the simple intensification and remission found in any one of the primary planets—and finally instead of the simple digression of the primary planets in latitude also a twofold digression.

What cause does the magnitude of the moon's eccentricity have?

In the *Harmonies* I demonstrate that the variety of lunar movements determines precisely the perfect fourth, which seems to have an affinity with the quadratures and syzygies of the moon. Therefore in order that this interval might be represented by a composite movement, the eccentricity was made to be as great as it is.

[612] *What diversity is found between those irregularities common to the moon and the planets and these irregularities proper to the moon?*

1. Just as, in the above, the movement of the moon around the Earth had two elements, as it were, one from the rotation of the Earth around its axis, the other from the joining [ex applicatione] of the solar light with this motor form from the Earth—whereof the first was free of the phases of the moon, and the other was bound up with the phases; so now also, of the two irregularities, that earlier one is found to be an accident of the earlier element or the mean movement and to have its own proper termini, which we shall call the apogee; and the first type of digression in latitude has its own termini, distinct from the termini of the apogee, and they are called limits and nodes. But the later irregularity, which arises from a later element, or else is an accident of the acceleration in the syzygies, and is called by Ptolemy the nodding [annutus] of the

epicycle, has termini in common with the lunar month and phases—as does the second type [forma] of digression in latitude.

2. That earlier irregularity both of longitude and latitude is always constant throughout all its periods, that is, it is forever of the same magnitude: but each of the later irregularities become greatest in one month only of any half year, smaller in the remaining months, and in certain months, which divide the year into two parts, almost nil, namely, where the opposite affects, of this second acceleration and retardation and likewise of the northern and southern latitude, begin to pass into the contrary halves of the lunations.

3. And so those earlier irregularities have their magnitude and the laws of their distribution from their own proper causes; but the second irregularities take their magnitudes and affects from the presence of the first irregularities in any one semicircle of lunation; they have laws of distribution alone separate, and adjusted to the cycles of lunations, but none the less similar to the earlier [laws].

4. The following fact is also related: that in the moon we find the movement of the apsides eastwards, and the movement of the limits [613] westwards, much faster than in the primary planets, not merely in the ratio of the faster periodic return of the moon, but quite perceptibly; and moreover the retrogression of the limits is more than twice as slow as the progression of the apsides.

The moon is not seen alternately to turn now this part of its body, now the opposite, to the Earth. For we always view the same spots on the face of the moon. Wherefore the causes of the approach and withdrawal of the moon from the Earth cannot be sought from this.

1. It is not necessary that at two opposite times of a period the lunar magnetic threads should point in a straight line towards the Earth; it is sufficient if at those moments they at least be deflected towards the Earth in their alternate regions, and if the posture of the threads remain parallel throughout the whole circuit of the moon. For it can come about in this way too that now one region of threads deflects more to the Earth, and now the opposite region. But if this deflection is small, our eyesight is not so sharp that in the disc of the moon it can observe very precisely whether on the rims of the lunar globe, which look towards the poles of the ecliptic, some minute particles present themselves to view, which are not seen at any other time. For those parts of the globe are shelving and of a very tenuous appearance, and the illumination frequently fails, now on this rim, now on that, on account of the inconstancy of the lunar countenance.

2. For a long time we left it as uncertain, whether there is a globe within the globe, like the kernel within the shell, having a different rotation from that—as the case of the Earth and also the movement in latitude suggested. And so such an inner globe could direct towards the Earth regions which are alternately reversed—notwithstanding that the outer crust always turns the same spots towards the Earth. For between these and similar things, it is uncertain exactly what the manner of this motion is: it is alone very certain that, whatever the manner is, it has been fitted to physical and magnetic causes, [614] *i.e.*, corporeal and thus geometrical; and I have put forward examples of such causes in both cases here.

*Then does not this second irregularity of longitude really come from some second
eccentricity or digression of the moon from the Earth, just as the first irregularity
has its cause in the change of the interval?*

No: the observation of the parallaxes of the moon, together with a consideration of the eclipses, opposes this; and the ratios of the distance of the
bodies from the first body militate against it—the ratios put forward in the
first part of this book. But it can be argued that there is absolutely no change
of interval bound up with the phases; because, though different experimenters
make different corrections in the case of this hypothesis, the magnitude of
this change is always made smaller and smaller. Ptolemy set it down as very
large; Regiomontanus reduced it; Copernicus halved it and transposed it
from the figure [*ex forma*] of an eccentric circle into the figure of the second
epicycle; again, Tycho Brahe got hold of it and claimed a part for the equant
circle, for which he together with Copernicus was accustomed to substitute
an epicycle of twofold movement. I have changed around the intervals at the
syzygies with those at the quadratures, and have transposed the cycles from
the month to the year; relying upon these discoveries made in latter times, I
at least found that absolutely no change of intervals occurred through the
cycles of the phases.

*Then from what comes this second acceleration and retardation, which are bound
up with the phases?*

From the varying relation of the eccentric circle of the moon to the phases.
For while the moon goes around the Earth, its mover, according to the simple
and forever uniform law of eccentricity, just as any one of the primary planets
goes around the sun; it comes about by accident that at different times the
moon has different distances from its movement's other mover, which accelerates the moon in the syzygies. For if its longer interval from the Earth
occurs at the syzygies, where its greatest acceleration takes place; then the
terrestrial form, unrolled in a wider sphere, is weakened at one of the syzygies—
weakened not merely in its own native and archetypal vigour, but also in the
reception of strengthening from the sun. In turn, if this longer interval between the moon and the Earth is found at the quadratures, where there is
no acceleration; then there is no loss in the reception of nil vigour, and no
gain with the short interval at the perigee.

In the diagram on page 952 are depicted in the globes of the Earth and
moon, the circles of illumination which divide the light part from the dark.
But since the apogee of the moon, through the whole year and thus through
all the positions of the lunar heaven, remains at the same point, *i.e.*, the threads
WF remain approximately parallel to themselves throughout the whole circuit,
but the Earth together with the lunar heaven crosses from point to point
hence the threads are joined in different ways at different times to the circles
of illumination—which spread out in accordance with a circle homocentric
with the sun and exhibiting the density of the light in longitude—as you
may see in the arcs *DT*, *EV*, *FW*, *GX*, *AO*, and *BP*. Accordingly the
same thing takes place also at the apogee and perigee of the moon, as
they always point towards the places 90° distant from the place or region of
point *A*, *B*, *Q*, etc.

But what if the longer interval stretches towards the sun? Will not the movement be weakened in that way? And yet the moon is passing through a denser light.

Indeed, this is what we remarked about above. For the light of the sun does not move [the moon] by itself but by means of the form of the terrestrial body, to which it transmits the laws and modes of its own work. Accordingly just as above light did not give the direction of movement, but the form of the terrestrial body gave a direction, somewhere plainly contrary to the direction in which the sun moves around its axis; so now also the motor form from the Earth is strengthened in proportion to its own native strength, weakly where it is weak, namely at its farther distance from the Earth its source; strongly where it is strong, at its shorter distance from the Earth—whatever the variation in the distance of the moon from the sun may be, [616] as above, in the case of the causes of the variation, we spoke of the counterbalancings of this.

What is the measure [modus] *of this monthly additosubtractive difference, when it is very great, and what is the cause of its measure?*

Tycho Brahe makes it equal to the physical part of the analyzed periodic additosubtractive difference [*aequationis periodicae solutae*], according to my formulation, because since the total periodic difference is approximately 5° I claim half of that for the physical cause, the customary amount for all the planets—that is, 2°30': accordingly Brahe exhibits the synodic difference to be that much—as if the motor form from the terrestrial body became precisely twice as strong at the near distance, and twice as weak at the far, through this strengthening by the light, and that is the time when it is without this strengthening. And if it is asked why that ratio holds, it seems that it can have no cause [617] except this relation of equality, as the most simple and therefore the most beautiful ratio.

But Ptolemy exhibits its measure as slightly greater, and just as much as above, from the addition of 132° 45' to 12 synods, we inferred the variation of one quadrant to be, namely 2°41'. But if this measure and magnitude is to be kept at both distances, then it seems that the cause must be transferred from design to geometrical necessity, namely, because the increase of the interval, *i.e.*, the eccentricity, utterly exhausts that which the acceleration by the light at that syzygy had given; but in turn at the other syzygy the subtraction of the eccentricity from the interval adds as much to the speed as was also produced by the acceleration given by the light.

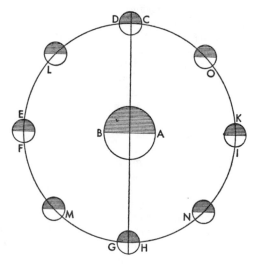

So in a month, which is without the synodic additosubtractive difference, namely when in this diagram the apogee is at *EF* and the perigee at *IK*, the parts of acceleration given by the light are equal at both syzygies, because the intervals between the moon and Earth at both syzygies, as *AC* and *AH*, are equal during that month; in the succeeding months, as inequality of intervals arises gradually at the syzygies— for example, if *L* were the apogee and *N* the perigee, then the distance *AC* would certainly be greater than *AH*—there also arises some monthly additosubtractive difference, which always becomes greatest during that whole month at *EF* and *IK*; finally in that month wherein the synodic additosubtractive difference is complete—for example, if the apogee is at *CD*—there is no acceleration at the syzygy *CD*, but at the other syzygy *GH* the acceleration is twice as great as in the month just spoken of; at that time at the quadratures *F* and *K* the additosubtractive difference is the greatest of all those which can occur during the total year. But at *G* very near to the perigee, there are four parts to the very small additosubtractive difference, (1) the optical, as in the planets, (2) the physical, as in the planets, (3) the variation due to light, (4) the intensification of the same on account of the distance being diminished. The ratio of these four parts to one another is of geometrical necessity. But the sum composed thus of all four was so attuned by design that *GH* the perigeal movement of the moon at the syzygy would be to *CD* the movement of the moon at the quadrature as 4 is to 3; and that the harmony would be the perfect fourth.

[618] For that reason it comes about that although these two, (1) the measure of acceleration at the syzygies and (2) the measure of eccentricity, are bound to one another by no necessity; nevertheless the eccentricity exactly destroys the acceleration at the apogeal syzygy and doubles it at the perigeal syzygy. I say that at present, while I have been unable to investigate it.

What probable cause do you assign to the so great speed of the apsides and limits of the moon, if the apsides and limits of the primary planets are incomparably slower?

Indeed the effect of the composition of motor virtues in the moon is made clear. For just as in the above pages we said that the simple force of the Earth was attuned to harmonic numbers—at any rate in the rotation of the terrestrial body around its axis, to 360 perfect days, but in the revolution of the moon around the Earth, to exactly the 12 months in one year, or in one periodic return of the centre of the Earth around the sun; so now too let us say that in the deflection and restitution of the threads of the moon—those by which the libration and those by which digressions in latitude are produced—their simple forces are attuned to the great length of the periodic time of the moon in the same proportion as was kept in the other planets. But just as in the above, on account of additional aid from the sun, both in making the terrestrial globe rotate and in making the moon revolve, the archetypal numbers were disturbed in their final effect, so that instead of 360 days there were 365¼, and instead of 12 lunations in the year, approximately 12⅓; so now too on account of this increase in the same acceleration of the moon from the illumination of the sun, the moon comes to the mean latitudes of its circuit before the threads are deflected in the right measure; and so the thread looks towards the sun from a deeper place (*loco profundiori*) than a quarter [of the distance] from apsis [to apsis]. And I inculcated above that the transportation of the apsides took place when that occurred. But it is quite right that this transposition of the apsides should be

perceptible, because that increase [619] is perceptible, namely approximately
11°; none the less this transposition is smaller, *i.e.*, slightly more than 3° in a
month: (1) because those degrees [of increase] are for the most part added on at
the syzygies, but the threads are most deflected and counter-deflected at the
mean longitudes without reference to the syzygies; in such a way that the affects
of degrees equal in number are very distant from one another; and it is probable
that something still remains hidden at that node, and that by our ignorance of
it the movements of the moon have not yet been worked out to the minute,
not even in Tycho's calculation. (2) Because the deflection of the threads is
anticipated not merely in place and time but also in magnitude for that very
reason. For if the moon had advanced more slowly, or if the deflection of the
threads were as great in the accelerated moon as it would have been in the slow
moon, the apsides would be transported even farther. But it comes about by
the acceleration of the moon that the thread points towards the sun before it
reaches the just measure of deflection originally assigned to it. And by the mixture of these things it comes about that a certain mean, $3\frac{1}{4}°$ between the nothing or something imperceptible—which would be the case without the acceleration of the moon—and the 11°—which the acceleration causes—overflows into
the movement of the apsides. The same things are to be said concerning the
impulsion or deflection of the threads of latitude; for it ought to have been imperceptible, as in the primary planets—if the moon, like the primary planets
was going to advance according to a simple force. But because the acceleratory
force which is added to the moon is evaluated at approximately 11° of longitude
in efficacy; and if this force fell upon these threads of latitude during the whole
cycle, it would deflect them, as not fortified against itself, the whole 11°; but
since it gets possession of them only at the limits where they are in opposition
to it, it gives them however a deflection of $1\frac{1}{2}°$ in one period. And the precession
of the limits follows upon this deflection.

But although from the observation of so many ages exact determinations have
been made of the magnitudes and ratio of these two movements of the apsides
and the limits; there still remains room for talent. For he who brings forward
such causes of these things that from the causes the magnitude itself follows,
will drive his chariot to the goal. And philosophers ought all the more [620] to
strive, because, over and above so many other experiments, in this question too
the moon is our mistress in acquiring knowledge of the celestial bodies, and by
its own example it throws light upon the nature of all the planets.

*How is it that besides its accustomed periodic latitudes the moon also makes synodic
digressions to the north and to the south?*

Just as that force of light which strengthens the terrestrial form, the mover of
the moon, borrows the region of movement and the proportions of the work
itself from that very form which is strengthens; moreover just as, in the case of
longitude, that force passes over into the inborn character [*ingenium*] of the
orbit according to the measure of the mutual joining; se we must set down that
that force does the same thing in the case of latitude. It facilitated the movement in longitude, because it extends in longitude; therefore it will facilitate
the movement in latitude, because it has the other dimension of latitude also,
i.e., because light is a surface partaking of density, as we have often recounted
from optics. Accordingly at the syzygies, where the thread of latitude is tangent

to the orbit and has been deflected in proportion to the latitude of the terrestrial form, the latitude of light joining itself to the terrestrial form facilitates the digressions, so that they take place at an angle greater than that which the thread makes with the plane of the ecliptic; and in that way the moon arrives in the quadratures at farther limits to the north and to the south than those which the thread showed at the syzygies. Conversely, during the other quarter of the year the thread of latitude, which is tangent to the orbit in the quadratures, does not fit itself to the dimension of the form of light in latitude, but points approximately towards the sun, just as the orbit itself of the moon does. Therefore just as in that place the movement in longitude is not at all facilitated by the light, but is made more difficult as it were; so the same thing overflows into the digression in latitude, so that it does not become greater than the angle by which the thread of latitude is deflected towards the ecliptic; and thus the moon does not arrive in the syzygies at limits farther than those which the thread showed in the quadratures. But what happens to the moon turning at the limits, or with what face the light of the sun beholds the moon, namely, when [621] the thread of latitude points towards the Earth, has nothing to do with latitude.

Since you reduce everything to the bodily threads of the globes and to the immaterial forms from the rotating bodies of the sun and Earth, and lastly to the light of the sun as a strengthening cause, you leave nothing to the animal faculties: hence you seem to be philosophizing just as if someone were to contend that its triple threads were sufficient to the belly for its operation and that there is no need of the animal faculty?

Rather, I admit a soul in the body of the sun as the overseer of the rotation of the sun and as the superintendent of the movement of the whole world. Nor in Book I did I deny absolutely that the planetary bodies had single souls as overseers of the rotation of their bodies. But just as it was not necessary to introduce a special soul into the threads of the belly; for it is sufficeint for one common soul from the heart or liver to advance, through its own form or through heat, into the belly and to employ the faculties of its threads; so too in the world that form—of light, or heat, and thus too, if you will—from the soul of the sun, flowing out together with light and heat and penetrating even where light and heat are shut out, *i.e.*, into the inner threads of bodies, seems to be sufficient; hence, just as the soul in the body has no power without the organ of the belly, so too the soul of the world has no power without these laws and without the geometrical lay-out of bodies.

Therefore let the status of the controversy be noted: for it is certainly one thing to reduce every cause [*rationem*] of dispensing celestial movement—although involving contradictions, and hence impossible—simply to the hidden forces of some soul, after rejecting all bodily organs and all the means which the human mind can devise—and that is the sanctuary [622] of all ignorance, the death of all philosophy, but is nevertheless the common practice of most of those who write or speak about astronomy, and in the above was noted in part even in Ptolemy himself. It is something else first to discern within the bodies everything suited for movement, so that the possibility of movements may be apparent—even with common examples—and afterwards finally to pour in a motor soul on top of all those things, as upon the human body compacted of all muscles and nerves. For if the soul can perform its functions anywhere by bodily organs, it will have no need in them for design and discourse, acts proper to an intellectual soul—just as, on the contrary, if it performed everything by design and discourse, it would not want those bodily organs.

In brief, the philosophers have commented upon the intelligences, which draw forth the celestial movements out of themselves as out of a commentary, which employ consent, will, love, self-understanding, and lastly command; the soul or motor souls of mine are of a lower family and bring in only an impetus—as if a certain matter of movement—by a uniform contention of forces, without the work of mind. But they find the laws, or figure, of their movements in their own bodies, which have been conformed to Mind—not their own but the Creator's— in the very beginning of the world and attuned to effecting such movements.

END OF BOOK FOUR, THE FIRST BOOK ON THE DOCTRINE OF THE
SCHEMATA OR OF CELESTIAL PHYSICS

BOOK FIVE

LETTER OF DEDICATION

[631] To the Very Reverend, Most Illustrious, Most Highborn, Most Noble, Vigorous, etc. Lords, To the Orders of the Archduchy of Austria-on-the-Anisana, To My Most Gracious Lords:

Four years after the publication of the first part of *Copernican Astronomy*, which contains the doctrine of the sphere unfolded in three books, one year after the publication of Book IV, wherein I have handed on the celestial physics, or the principles of the doctrine of the schemata of the planetary movements; there at last follows some time the speculative part—so called from the schemata, that is, the manual instruments wherein as in mirrors [*speculis*] the movements of the single planets are represented.

[632] If I regard the circumstances of the time, this publication, alas! arrives in town late, after a very destructive war has arisen and the assemblages of students for whom these things are written have either been dispersed by the confusions of war or thinned out and wasted away by the expectation of war; after Austria, hitherto my nurse and benefactor, has struck against a very hard reef and seems to be called away from the guardianship of these beauties to serious care for her own safety; and after I too, forced by the hatefulness of a private enemy to leave my home at Linz, have been moving around for nearly a year away from home.

If the causes of such great delays have to be mentioned, I shall not allege the supineness of my publisher—which has lasted from the publication of the doctrine of the sphere up to now—or the inconveniences of present war or the fears of threatening war: the fact that this publication has been prevented till now calls for thanks, not blame. Then what cause shall I mention, whereby I may preserve my reputation and wash away the charge of negligence? "While our womanly manners are at work," says Comicus, "while they dress their hair, a year passes by." But if the manners of astronomy are known to anyone, he will be able to say that he has never known a slower or more painstaking woman. [633] For unless this time had been interposed, wherein my designs might reach their maturity, there was danger that that finicalness, now that the whole world was squeamish, would demand new outlays and new adornments. For the computation of the *Ephemerides* and the publication of the books of the *Harmonies*—the works of the intermediate time—gave me many warnings that although most of the things which have to do with the six planets had been drawn up or at least indicated twelve years ago in my *Commentaries on Mars*: and although, taken over from that and put together into the form of a textbook, they had remained in my writing desk for seven years now, awaiting the work of the printer and engraver;

nevertheless as often as I reread them, what with additions or elucidations or transpositions of the text, the necessity of making a new copy was imposed upon me. To such an extent that not a trace of the first draft was left in what was shown to the printer. Now as regards the moon, the last of the planets: when I first gave my attention to the publication of this *Epitome*, I had no special concern about the moon, because Tycho Brahe's hypotheses about the moon already existed, and they could [634] in general be found to be equivalent to those hypotheses too by which the manifold movements of this planet were reduced to my physical causes. Moreover those hypotheses existed as shadowed forth in the *Commentaries on Mars*, and as worked out further in my *Hipparchus*. But they were of such sort as to suppose two circles in the moon, eccentric in each case—a thing quite inimical to physical speculations, and hence intolerable. The computation of the *Ephemerides* rested upon those bases: and from the foreword to it, it is apparent that the form of the calculation has been changed again and again, because its agreement with appearance fluctuates and vacillates everywhere.

Finally the great felicity of my speculations freed astronomy from this cross in the month of April 1620, when, after the physical causes had been considered more carefully, it appeared that the second eccentric circle of the moon was superfluous, so that there was no further need for imagining it with reference to the movements in longitude. And it was already time to put the finishing touch upon the fourth book of the *Epitome*, which is about the principles of the doctrine of the schemata. That done, I transferred my attention to the publication of it, in the midst of the Bavarian armies and the frequent sicknesses and deaths both of soldiers [635] and of civilians. However, in the year 1621 the *Ephemeris* was at once computed, and—after the fashion of my other *Ephemerides*—the foreword was made to signify publicly my rejoicing over the conquering of the second eccentric circle of the moon. But, forestalled by the necessity of my journey, I could not yet publish this *Ephemeris*.

Now as regards this last part of the *Epitome*, comprehended in three books; although after the publication of Book IV, I am away from home and spend no little time in journeys and in the law-courts, nevertheless I have been allowed to be at leisure the greater part of the time, and I have devoted all that time to the care of this publication. When I came to Tübingen at the end of the year 1620, in order to expound to Maestlin a new rationale of lunar hypotheses, I began to write down questions about the other planets and the moon too in accordance with the physical hypothesis finally discovered.

As soon as I returned to my family at Ratisbon, I reread the questions and gave them over to be copied. In the meantime I turned to the last part of Book VI, up to now postponed, because I hoped of its being easy, and it seemed that it could be put together between proof-readings; but [636] I found it laborious, not so much for the difficulty as for the multitude and variety of questions and the worry over the method. A short time I spent at a monastery passed in drawing up the ancient epochs and in computing the eclipses. At once, as soon as I returned to Tübingen, I saw that the fourth part of Book VI, on the moon, would have to be reshaped and work repeated, because the definitions conceived verbally did not yet accurately represent the force of my hypothesis.

During the last months May and June, Stuttgart gave the last little book, which was also included in the last part of my cares up to now: because the astronomers have too little evidence concerning the movements of the eighth sphere; but most of the things which could be said concerning this matter had been conceived by me in the *Commentaries on Mars*, in Book III of the *Epitome*, long published, and on other pages. Nevertheless many things arose on the occasion of my conversation with Maestlin, my old leader in taking the route of Copernican astronomy, and many things through the reading of books which I had hitherto been unable to get hold of in Austria. And if publication had not been postponed till now, these things would necessarily have to have been passed over.

[637] Meanwhile, with the shore of this voyage in sight, *i.e.*, the end of this work, and after being refreshed by money sent secretly to Linz, then by the argument of the faith and kindness of your Very Reverend Doctor of Divinity, Antony, President at Krembsmunster, and finally with a truce intervening in the law-courts—my great grief—I gave June over to the journey to Frankfort, and to looking after the printing. And here once more, while the work is being undertaken, while the pages, diagrams, and forms are being prepared, a month passes by. And this sidereal lady, who heretofore bore witness to her fretfulness by her countenance and her noddings, now ratifies and exercises her fretfulness, after finally coming to the press, by quarrels and abusive language, only not by hands and weapons.

Therefore, Very Reverend, Illustrious, Highborn Lords, let her stand before you to plead my cause, which can be got hold of for me from the daily delayings of this publication. Come to an understanding with her. If you have become experienced in listening to her sarcasm, you will not easily demand scrupulous accounts of time from him who proves that the case is with him as with her, especially if he can show the value of the time and the work.

But after I myself too, whom I believe to care for those arts, *i.e.*, the proclaiming of the divine works, [638] and who has followed the footprints of divine providence in an indefatigable search—after I recall to mind what utility the little book will have drained from this delay in time; I am not so frightened by your adversities, Masters—which have meanwhile risen up against you and the wretched province or else seem to make further threats—as not to perform my task, fulfill the promise made to you in the dedication of the little book on the sphere, and pay my debt, because I have hitherto been living off your subsidy. For I hope that there is so much divine mercy left in the store-house that He will that this horrible tempest be calmed, the clouds dispersed, and the sun at last shine again upon the penitent, and that some place be left even in Austria for these peaceful arts, the labour over which He does not stop caring for, and that there be gathered together in Austria once again some number of those who learn from these arts the praises of God their Creator. I hope that this little book will be of service to those people. For it contains the first adumbration, so to speak, of the Rudolphine tables, and the approximate numbers. If these numbers are assumed as true, the lovers of this discipline may exercise themselves with them until the Rudolphine tables themselves come out, furnished with everything accurately corrected, fixed for calculation, and ready for use. Furthermore, [639] if outsiders get any utility from my books, as there are very many people not only in Germany but also in the surrounding kingdoms and provinces who seek these books at Frankfort; it is right for them to learn from this dedication of mine that, for whatever value this book has, they ought to give thanks also to your liberality, Masters, wherewith uninterruptedly you have fostered me throughout these very difficult times. With that learnt, according as each is most disposed towards the arts of mathematics, most devoted to God, and most assiduous in gratitude, the crown of the virtues; so let him very frequently join his prayers with mine to the most merciful God: that, with the tumults of war calmed, the devastation restored, and hatred extinguished, golden peace returning may smile upon the Empire of the Most Powerful Divine Ferdinand II, Emperor of Rome, Our August Lord, and that God may revivify all the provinces of His Majesty, and especially Austria-on-the-Anisana, with the fruitful showers of His grace. And that finally to you, Very Reverend, Illustrious, Highborn, Noble, Vigorous Masters, God may bring and make permanent for many years safety, health, riches, and dignities, for His own glory, for the preservation of the Church, for the adornment of the most glorious Emperor, and finally for the necessary cultivation of those arts [640] wherein the divine name is held in honour. Farewell, Masters, and hold within your liking your little client, who is absent in body for a short while, but quite present in spirit for all acts of obedience. At Frankfort on the Kalends of July 1621.

Reverend and Illustrious Lords, Farewell,
your most devoted Mathematician,
JOHANNES KEPLER

SECOND BOOK ON THE DOCTRINE OF THE SCHEMATA

PART I. ON THE ECCENTRIC CIRCLES, OR SCHEMATA OF THE PLANETS

[641] *If you set up no solid spheres in the heavens and if all the movements of the planets are regulated by natural faculties, which are implanted in the bodies of the planets: then I ask what will the theory* [ratio] *of astronomy be? For it seems that the theory cannot do without the imagining of circles and spheres.*

It can easily do without the useless furniture of fictitious circles and spheres. But there is such great need of imagining the true figures, in which the routes of the planets are arranged, that we are impoverishing Astronomy and that the big job to be worked on by the true astronomer is to demonstrate from observations what figures the planetary orbits possess; and to devise such hypotheses, or physical principles, [642] as can be used to demonstrate the figures which are in accord with the deductions made from observations. Therefore when once the figure of the planetary orbit has been established, then will come the second and more popular exercise of the astronomer: to formulate, and to give the rules of, an astronomical calculus in accordance with this true figure, or even to make use of the figure as expressed in material instruments not otherwise than the solid spheres of the ancients were used, and through these figures to lay the movements of the planets before the eyes.

Therefore what is the subject-matter of the Fifth Book, the second book on the doctrine of the schemata, and how do you keep its subject-matter separate from that of the preceding Fourth and the following Sixth?

So far, in the Fourth Book, the physical principles of movements—among other things—have been demonstrated by reasons and by experiments. Out of these physical principles the Fifth Book will form the figures of the planetary orbits and will explain the powers [*potestates*] of those figures; and there the inmost sanctuaries of geometry will have to be searched. The Sixth will teach the use of these figures in the schemata of the single planets and put them into operation. So the Fourth contains the theory; the Fifth, the instrument; and the Sixth the practice; the Fourth was physical, the Fifth is geometrical, and the Sixth will be properly astronomical.

How many parts does Book V have?

Two: in the first part the eccentric circle and its plane are connected up with physical causes: in the second, there are given definitions of the astronomical terms which occur universally among all the planets in the case of the eccentric circle; and the method of calculation is explained in so far as it relates to this part.

Then what sort of figure of the planetary orbit is formed according to the physical principles of Book IV?

If the body of the planet did not have magnetic threads, [643] so that in accordance with one of the regions of magnetic threads it is drawn to the north, and in accordance with the other region towards the south—in accordance with one region it is attracted towards the sun, and in accordance with the other is repelled; then the sun in the rotation of its body around its axis, carrying around the immaterial form of its own body throughout the widest spaces of the world, would carry around with itself the planet held fast by that form, and (1) if the planet at the start had been situated under the ecliptic, its whole route would be arranged along the plane of the ecliptic; (2) and so the planet would always return to that very point where the start was made; (3) the body of the sun and the planetary orbit would have the same centre; (4) the figure of the orbit would be a perfect circle; (5) the planet in all equal portions of this circle would be moved with equal speed.

But because we have laid down that in the body of any planet there are present twofold threads:therefore by the mingling of the faculties of the planet's body and the sun's motor power, (1) the planet describes an orbit oblique to the ecliptic; and because the threads of latitude remain in approximately a parallel posture during the whole circuit but not wholly so, and hence they are deflected gradually after many revolutions: therefore (2) the plane contained by the orbit of the planet is approximately a perfect plane but not wholly so; and hence, when one revolution has been completed, the centre of the planetary globe does not return exactly to its starting-point, but it entwines a new circle with the first circle traversed and described—after the fashion of the circles of the natural days—of which I have spoken in Book III, page 291—or after the fashion of the thread which the silk-worm drops, throwing the thread around itself and building a little house by the inweaving of many entwined circles. For this reason the longest digressions in latitude are not found in all ages in the same parts of the zodiac. And because the threads of libration make the planet to be drawn away from the sun to one side but then to be driven away from that region; accordingly the planet (3) describes its orbit around the sun but not as around its own centres, that is to say, it describes an orbit eccentric to the sun: [644] and for this reason the orbit is not (4) a perfect circle, but one slightly narrower and

more pressed in on the sides, like the figure of an ellipse. (5) For the same cause
and because the form of the solar body, which form gives movement to the
planet, is slighter and weaker in the larger circle, it is not possible for the planet
to be moved with the same speed in all the parts of its orbit; but the planet is
slow at a long distance from the sun and fast at a short distance. Finally because
too the threads of libration are moved out of their parallel posture by very many
successive revolutions, so too the positions on the zodiac where the planets are
highest and slowest do not always remain but gradually move eastward.

*Have you not described an intricate figure for the route of the planet and one that
cannot readily be laid before the eyes especially in a plane?*

Even if this is true, it is nevertheless not new to astronomy or something
found only in Copernicus; and there is no need for all things to be represented
in the same plane; but those interweavings, which have arisen from the very
slow movement of the boundary-posts of latitude and altitude, can be disen-
tangled with the same dexterity which the ancient astronomers employed and
with less apparatus.

How did the ancients disentangle the movements of latitude and of altitude?

They devised for the latitudes one circle or sphere which is the carrier or de-
ferent of the nodes, the outermost circle of the whole schema of the planet—
and for the altitudes two spheres of unequal thickness, which they named the
altitudes of the deferents [*deferentium auges*].

Why do you judge that those spheres or circles should not be used?

Because they were made for laying physical explanations of movement before
the imagination rather than astronomical. And so by their use [645] those false
physical opinions concerning the solidity of the circles or spheres were established; and in turn the true judgments were obscured concerning the causes of
their irregularities and of their very slow transportation—these causes were
demonstrated in Book IV.

*Therefore what do you substitute for these three circles or spheres in laying your
astronomical explanations before the imagination?*

It is sufficient if we draw two straight lines from the centre of the sun, one
through the intersection of the planet's orbit with the ecliptic, the other through
the centre of the planet's orbit—both lines, on either side and everywhere under
the fixed stars—and if we teach that the movement of the first line is under the
ecliptic towards the westward signs and that the movement of the second line
is eastward under the circle, in the sphere of the fixed stars, which is in the same
plane as the orbit of the planet—both movements most regular, the first movement from the mean equinoctial point and the second movement from the line
of intersections [of the planet's orbit with the ecliptic]. Unless something should
here be taken from Book VII on the ground that even the ecliptic is dislocated
and does not always run through the same fixed stars.

*When this separation has been made, what remains for our imagination concerning
the figure of the route of the planet?*

The orbit remains as a perfect ellipse in a true and most regular plane inclined
to the plane of the ecliptic at constant angles, by which plane of the ecliptic

this orbit is cut in the line drawn through the centre of the solar body—as was stated on page 946, Book IV. The planet moves in this orbit with unequal speed through its parts and returns to the intersections, and so to the equinoctial points—not to say to the fixed stars and to the line through the centres, by absolutely equal measures of periodic times, so far as they are considered in themselves.

Does not this imaging trespass upon the physical causes and measures of the movements of one period?

Not very far, provided we keep in mind that those things which are abstracted by means of the said two lines from the real inweaving and interconnection of many orbits [646] are not furnished physically by those two lines but by the inclination of the real threads of the planet's body.

By what right do you make this also a part of Copernican astronomy, since that author abided by the opinion of the ancients concerning perfect circles?

I admit that this formulation of the hypotheses is not Copernican. But because the part concerning the eccentric circle is subordinate to the general hypothesis which employs the annual movement of the Earth and the stillness of the sun: therefore the name comes from the more important part of the hypothesis. Moreover, this small part of the hypothesis is bound up with necessary arguments arising from the repose of the sun and the movement of the Earth, the doctrines of Copernicus; and so this part has a good title for being referred to Copernicus.

What is the method of procedure whereby it will be demonstrated by the physical causes established in Book IV that such a figure of the orbit arises and what speed the planet has through the parts of its orbit?

We must start with the approach and withdrawal of the planet from the sun; and first, we must determine the geometrical measure of the strength of the forces exerted in giving the planet a movement of libration in any posture of the threads. And secondly, we must prepare a compendious geometrical measure of the effect of the attraction or repulsion, which is heaped up through any whole arc of the orbit by all the increments of the forces. Thirdly, we must demonstrate that the elliptical figure of the orbit arises from such a libration completed during the revolution. Fourthly, we must show that the plane of the ellipse exhibits the measures of the time and of the delays which the planet makes in any arc of its elliptical orbit. Fifthly, we must teach the equivalence of the plane of a circle and the plane of an ellipse with respect to this measuring of time. Finally, we shall have to demonstrate that when the revolution of the threads of latitude has been thus procured, as [647] was laid down in Book IV, the regularity of the orbital plane is established. By these demonstrations the studious astonomer—for the popular astronomer has no need of Book IV or of this first part of Book V—will have his curiosity satisfied concerning this part of the calculus of movements. The other part of Book V will teach how to construct this calculus; and Book VI, how to bring it into use by the application of this elliptical orbit and its plane to the great circle or sphere.

1. Concerning the Increment of Libration

Begin with the first and say on what principles the measure (modus) *of increment of the libration in every position of the planet is formed or determined.*

Two causes, one active and one passive, come together for the formation of this increment. The active cause is the quantity (*modulus*) of the forces of libration in themselves, that is to say, how great the quantity (*modulus*) is found to be in any equal arc of the eccentric orbit. The passive is the varying lay-out of the planetary body with respect to the sun; for not every lay-out receives or admits that whole quantity (*modulus*) of forces, but any lay-out receives its own proper portion [of the forces].

So what measures the quantity of the forces for giving the planet a movement of libration?

These three do: first, the distance of the arc of the orbit from the sun; secondly, the magnitude of this arc; and thirdly, the time which the planet takes to turn through this arc.

What does this distance of the arc—and of the planet in the arc—from the sun have to do with the forces of libration?

The ratio of the distances is the inverse of the ratio of the tenuity of the solar form. This same form carries the planet around and gives it a movement of libration, now attracting it and now repulsing it, as was said in Book IV, page 903. And so the farther the arc is distant from the sun, the weaker at any moment of time is the libration of the planet moving in the arc. For this reason alone, the sun would expend unequal forces in different arcs of the eccentric circle which are equal to one another.

[648] *What effect does the magnitude of the arc of the planet have?*

Because in a long arc much force is expended, and in a short arc little force: therefore if equal arcs are assumed, so much and so far should the forces be equal.

What does the time, taken separately, have to do with the increase of forces, and what do all three causes, taken together?

Since—Book IV, pages 63-66—it has been shown that the farther the planet is distant from the sun, the longer does it delay in the equal arcs of the orbit; and so much the longer does it feel the motor force of the sun, the greater the magnitude of that arc: but since it has already been said that the farther any one of the equal arcs of the orbit is distant from the sun, so much the feebler is the movement of libration which the planet undergoes: wherefore the more feeble the libration is in one moment of time, in any of the equal arcs of the orbit, so much the longer does the planet revolve and undergo a libration in that arc. Therefore since the feebleness of the forces is balanced by the stretch of time wherein the planet experiences those forces in itself—and this occurs in the same ratio on both sides, namely, in the ratio of the distances from the sun: hence there is the final effect that in the equal arcs of the eccentric circle the quantity (*modulus*) of the forces of libration is disbursed by the sun and is absolutely equal with respect to the sun as an agent. See diagrams on pages 63 and 94.

Now therefore state the measure of the portion of the quantity of the solar forces which the planet admits into itself in any posture with respect to the sun.

We must consider the angle which the rays [*radii*] of the sun make with the magnetic threads of the planetary globe. For the sine of the angle complementary to this measures this portion of forces admitted. For since the efficient causes of the libration are the ray of the sun and the magnetic threads of the planet's body, two physical lines; it is right to seek the measure of the strength of the libration from the angle between these lines and from its sine.

[649] For example, let *A* be the sun; and *I*, *E*, the centre of the planetary body; *RP* the line drawn through *A* the sun and through *B* the centre of the orbit; *EG* and *IH* the magnetic threads will be practically perpendicular to *RP*—at least if the counterbalancing of the semicircles is considered—and *H* and *F* are the solipetal termini. For it was laid down in Book IV page 95 that during the revolution of the body the magnetic threads stay practically parallel to themselves and at points *P* and *R* provide no occasion of attraction or repulsion, because there the positions of both the solipetal and the solifugal termini are equally distant from the sun at *A*. But in the intermediate positions, where the solipetal or the solifugal termini regard the sun directly, the strength of the libration is greatest of all. *AE* and *AI* are the rays of the sun. Let lines *ED* and *IO* be drawn parallel to line *RP*, and let perpendiculars be dropped upon them from points *F* and *C*, wherein the rays of the sun cut the middle circles of the planetary globe; and let the perpendiculars be *CL* and *FK*. Here the angles made by the rays of the sun with the threads are *AEG* and *AIH*; the complementary angles are *CED* and *FIO*,or the arcs *CD* and *FO*, and *CL* and *FK* are the sines of these angles, according as *IH* or *EG* is the total sine of 100,000. Therefore it is established that as *EG* and *IH* are to *LC* and *KF*, so the total quantity of the forces from the sun which are present at *I* or *E* is to the portion which the planet admits at the postures of the threads *EG* and *IH*.

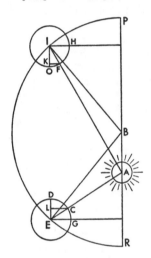

Why do you take the sine as the measure, rather than the complement of the angle or arc?

Because, although any magnetic thread is present in a spherical body, nevertheless it is not a circle, but is a physical straight line; because it works most strongly towards undergoing attraction or is very strongly disposed towards admitting the forces of the solar ray into itself, when it points towards the sun in a straight line: or else—what is the same thing—when it is perpendicular to the plane of illumination of the circle, which bounds the part of the globe which is turned towards the sun; but when [650] it is inclined obliquely to that plane, it is equivalent, as to a shorter line, to the perpendicular drawn from its terminus to that plane. Thus the solar ray, considered with respect to the work of heating, heats most strongly when it strikes a surface at right angles; but when at oblique angles, it warms less, in the measure that the line drawn from the sun perpendicularly to the same plane continued is shorter than the oblique ray.

The following explanation will be more elegant: If you consider that the whole globe is made up simply of threads, whereof the longest threads are those in the great circle of the globe; and the shorter, those in the latitudinal circles. In this way not only will *EG*, like *IH*, be a thread; but also the aforementioned sines *LC* and *KF*, which are marked out by the solar ray *AE* and *AI* at the termini *C* and *F*: those sines are latitudinal threads. Therefore the smaller *CL* and *FK* are than *GE* and *HI*, so much the less of the forces from the solar ray does any one thread of the whole body admit into itself, on account of the solar ray falling obliquely upon itself. Thus the solar ray, by marking out the latitudinal thread, marks out the sine, which is the measure of the portion of its virtue received into the threads.

Furthermore, every artificial or natural movement, in which the same or analogous principles concur, is measured out by the sines of the angles; but principally and most clearly, the movement or tendency [*nisus*] of the arms in the balance and in the lever. Accordingly since this libration too takes place between movements which are natural in a wider signification—because the libration power of the solar form is a partaker of dimensions, and somehow, though without matter, corporeal; but the lay-out of the threads in the planet again is corporeal; it is not absurd that this libration too should get the same laws as the balance and the lever. And that is all the more probable in the case of the libration towards the sun, because the progression of the planet in longitude along its orbit, by reason of intensification or remission, *i.e.*, by reason of speed and slowness, obeys the laws of the same balance or lever, as was said in Book IV, page 47 and 66, and will be made clear below in many ways.

[651] *Compare this speed of libration with the proportionalities of the balance.*

The line from the sun into the threads is like the haft of the balance, the threads like the arms of the balance; and the regions of the threads like the tray; and the weights in the trays are, in the case of the planet, the attraction towards the sun, or repulsion from the same; and both are of the same family of things. For as the sun attracts the planet, so the Earth attracts bodies, and on account of this attraction, bodies are said to be heavy. But the sun attracts the planet with respect to one region and repels it with respect to the other, and does this with varying intensity; while the Earth attracts weights without any distinction being made as to posture. Accordingly that which in the balance is inequality of weights, in the planet becomes the diversity of posture of the threads with respect to the sun: for here the planet exhibits both weights on the balance as the same weight. And just as in the balance, the heavier weight comes down towards the Earth, and the lighter moves up away from the Earth; so in this case the whole globe of the planet suffers the affect (*affectionem*) of the prevailing region. Hence if the friendly region is more attracted by the sun [than the unfriendly region is repelled], the whole planet approaches the sun; but if the unfriendly region is more repelled, the whole globe of the planet is repelled by the sun. Therefore too the measure, in accordance with which the weights of the balance are at war with one another, will be dominant in the disbursing of this attraction and repulsion. But in the balance the victory of the weights is evaluated [652] at the sine of the complement of the angle between the haft or handle (*manubrium*) and the arm of the lighter weight, as will be proved. Wherefore too in the libration of the planetary body towards the sun, the passion of the region of thread nearer the sun will overcome the passion of the opposite region in the ratio of the sine of the complement of the angle between the solar ray and the thread. But the effect of victory, in the movement of the planets, is the strength of the libration belonging to each position. Accordingly this strength, or the increment born of that [strength] of libration, will similarly be evaluated at the sine of the complement of the angle with the threads.

Let AD be the handle or haft; and let AB and AC, the arms in the same

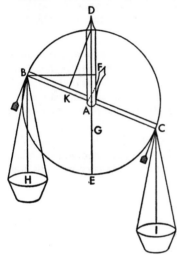

straight line BC, be each equal to AD. Let H be the lighter weight hanging from B: and I, the heavier weight, hanging from C. Accordingly the weights, which by their power (*potestate*) are at points B and C, have an altitude as great as BC the length of the arms; and the weights contend for this altitude, which is DE. For if the greater weight completely overcame the other, the arm BA would coincide with the handle DA; and the greater weight C would be in the place of altitude E, and would lift the lesser up to D the very top; but because the heavier weight does not overcome the lighter completely, therefore the perpendicular BF drawn from the end of arm B to the handle DA shows that the weight B is lifted through FA, a part of the altitude, and that the weight C goes down the same amount, namely through AG. Accordingly

$DF : FE : :$ weight H : weight I.

$FE : FG : :$ weight I : weight I − weight H.

$DE : FG : : DA : FA : :$ weights I & H : weight I − weight H.

But if BA is set down as the total sine, FA will be the sine of angle FBA, which is the complement of angle FAB.

In the same way if EA is the ray from the sun, BC the magnetic thread of the planetary body, H or B the lesser strength of repulsion, and I or C the greater strength of attraction—because C has approached nearer to the sun than B—then, if BA exhibits the strongest attraction and with no angle BAD, AF will represent the attraction at the existing angle BAF or GAC.

Apply these things to the proportionality of the lever too.

The proportionality [*ratio*] of the lever is the same, with only the following difference: [653] in the balance the fulcrum A is at the midpoint between B and C the extremities of the arms, and thence the unequal weights make BC not remain parallel to the horizon: but in the lever the line of weights remains parallel to the horizon, while the fulcrum divides the length of the arms not at the midpoint but nearer the heavier weight, so that the arms have the inverse ratio of the weights.

For example, if DA the handle of the balance is equal to the arms BA and AC, a lever will be formed, and the weights hanging from B and C will be suspended at equilibrium with respect to the horizon—as follows: DK the perpendicular drawn from D to BC will be the handle; and BK and KC will be the arms; and as formerly DF was to FE, so now BK is to KC. Then as BK the shorter arm is to KC the longer arm, so the lesser weight H to be suspended from C is to the greater weight I to be suspended from B.

The reader must be warned that mechanical experimentation is difficult: because the weight and thickness of the arms themselves cannot be guarded against mechanically. Now they ought to constitute geometrically a pure line without weight and width. It may be seen in Archimedes how we must meet this hindrance in part.

I grasp that the measure of the strength, or of the increment of libration, at any given posture of the threads of the planetary body must be sought from the complement of the angle made by the thread with the solar ray; but because it seems that this angle is discovered with difficulty, in that not only is the body continually transported from place to place, but also its threads are deflected; does not this measure seem uncertain and therefore unsuited for use?

On the contrary, on account of this deflection of the threads, that angle can be converted to the arc of the orbit, so that from this arc will come the same sine, *i.e.*, the same measure, and for this reason that measure is very well adapted to use.

Teach and demonstrate this converting of the angle to the orbit.

You remember that at the start when the planet is at the apsides, [654] *i.e.*, at the beginning of the orbit, the angle between the ray of the sun and the thread is a right angle. Again in Book IV, page 101, it was shown that thread NQ of this figure points towards A the sun, or is one with NA the ray from the sun—the angle here being null—when PN a quadrant of the orbit from apsis P has been traversed: hence the arc of the orbit from the apsis measures the complement of this angle. Therefore it remains to be demonstrated, that even the intermediate angles made by the thread and the sun, as HIA, between a right angle and no angle, have as their complements the middle arcs of the orbit, as PI, between no arc and a quadrant, so that the two together make 90°.

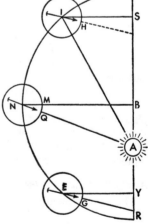

It is demonstrated as follows: On page 944, it was said that $IS : NB :: $ angle $HIS :$ angle QNB. That [ratio] is used for the sake of discerning IS and NB, although, by the force of physical speculation, that [ratio] is true of the sines of angles IAP and NAP. But now

For

$$\sin AIB : \sin ANB :: \sin IAP : \sin NAP.$$

And

$$BI : BA :: \sin BAI : \sin BIA.$$

$$BI : BA :: BN : BA :: \sin BAN : \sin BNA.$$

Therefore

sin BAI : sin BAN : : sin IAP : sin NAP : : sin AIB : sin ANB.

Therefore if the foregoing terms are compared with one another, angle HIS will be found equal to angle AIB, and angle QNB to angle ANB; and, if equals are subtracted [from equals], angle SIB will be equal to angle HIA—just as, by proportion, angle BNB will be equal to angle ANA. But the measure of angle SIB is arc IN, because the measure of angle SBI is arc PI. Therefore too the measure of angle HIA will be arc IN, the complement of arc PI. Therefore, given PI the arc of the orbit, SI the sine of the arc, i.e., the measure of the increment of libration, is also given immediately.

2. On the Sum of the Libration Gone Through With

I comprehend the measure of the increment, or of the strength of libration, at any given moment. But I should like to know the measure of the part of the libration gone through with from the beginning up to that moment.

That measure is got from the versed sine of the same orbital arc so far traversed. For as the whole major diameter of the ellipse is to the total libration, or—what amounts to the same—as the semidiameter of the orbit is to the eccentricity, so also the versed sine of each arc on the orbit starting from the apsis is to the part of the libration which is gone through with in the meantime, while the planet is traversing that arc.

By what means is this demonstrated?

By means of that very measure of the increments of libration which has just now been confirmed by its own demonstration.

For let PD be a perfect circle whose centre is B. Let A be the sun, and $PBAR$ the line of the apsides: and let P and R be the highest and lowest apsides. Let AB be the eccentricity; and let its double, PB, be the total libration. Now let the circle be divided into equal least parts—starting at P. And let the parts be PK, KG, GD, DN, NS and SR. And from these divisions let the perpendiculars KX, GF, DB, NA, and SY be drawn to PR.

Therefore, by the foregoing, as the sine KX is to sines GF, DB, NA, SY, and RR—a point instead of a line—so the increments of libration corresponding to arcs PK, KG, etc. are to one another—that is, as PM is to MI, IF, FQ, QV, and VB. And that is true provided it is understood that the division is continued to infinity, when KX and RR are understood to be equal. Therefore since points P, M, I, F, Q, V, and B are put down as separating the said increments of libration; let them be transposed to the different distances of the planet from the sun A: that is to say, with A as centre and with intervals AM, AI, AF, AQ, and AV let the arcs ML, IH, FE, QO, and VT be described in such fashion that a planetary orbit [656] which is an ellipse may be understood to drop from P to R through the points L, H, E, O, and T. The distances of the planet from the sun will be AP, AL, AH, AE, AO, AT, and AR. But the versed sines of the said arcs PK, PG, etc. will be PX, PF, PB, PA, PY, and PR. I say that as the whole diameter PR, the sagitta of arc PDR, is to PB the total libration, so the sagittae of the single arcs are to the single increments of libration, namely so is PX to PM, so PF to PI, so PB to PF, so PA to PQ, so PY to PV.

For it has been put down that PM and PI the parts of the libration are in the ratio of the sines KX, GF, etc. But now also the parts PX, PF, etc. of the total sagitta PR are in the same ratio of the sines KX, GF, etc.—and in the same state of infinite division, where—no less than before—the point R plays the role of line RR.

Therefore, by alternation, the parts of the libration correspond in the same ratio to the parts of the sagitta—and as a consequence, any total portion of the libration from the beginning P corresponds to its total sagitta in the same ratio.

[657] *Whence do we know that the parts* PX *and* XF *of the diameter* PR, *considered as the sagittae are in the ratio of the sines* KX *and* GF *which determine them?*

Pappus made a demonstration, *Mathematical Collections*, Book v, Prop. 36. If a sphere—to be understood by PGZ—is cut by any number whatsoever of parallel planes, as KW, GZ, etc., then the surface of the sphere and the axis of the sections, as PR, are always cut in the same ratio: so that, as the spherical surface KPW is to portion of the axis PX, so too the surface $KWZG$ is to the portion XF, and so for the rest.

But if the spherical surface is understood to be divided into an infinite number of equal zones, then any zone, say KW or GZ, will be as a circle without width. But the circles KXW and GFZ are to one another, with respect to length, as their semidiameters KX, and GF, etc. Wherefore too the corresponding portions of the axis PR, namely PX, XF, will obey the ratio of the sines KX, GF, by which they are determined.

See the *Commentaries on Mars*, Chapter 57, for an attempted demonstration of the same theorem by means of numbers and the sectioning of the circle. But in that place this ratio seemed to be somewhat deficient, because I had not yet read Pappus. But the reason for that was that I did not take the first sagitta from a small enough arc, and that is just as if, in Pappus, you were to divide the spherical surface into parts not smaller than 1° in width. For then the width of the narrowest zone would necessarily be twice that which would be true.

Although PK, KG, *and the remaining arcs taken from the circle are equal, yet it does not seem that* PL, LH, *etc., the arcs of the true orbit are equal, but are greater towards* E. *Does not this destroy the certitude of the demonstration?*

Not at all. For the fact that the arcs are greater towards E [658] must be attributed to these very librations, as will be apparent below. But the same thing cannot be either the only cause or a concurrent cause of itself: so that I may pass over the disturbance, because, even if any is to be admitted, it would clearly be imperceptible.

3. On the Figure of the Orbit

I see that the measure of the libration is found in the versed sines of the arcs of the orbit starting from the apsis—by the principles and causes of movement which have been assumed. It is left for you to prove that by this form of libration an elliptic orbit is constituted, concerning which you have said the observations bear witness.

That the planetary orbit *PLHEOTR* and the opposite half becomes an ellipse is demonstrated from the properties of this figure which are identical with the properties which the libration so far described exhibits.

What are the identical properties of the ellipse?

1. It is clear, from the *Conics* of Apollonius of Perga, that the ellipse [659] around which a circle is circumscribed, with the longer diameter of the ellipse as the common diameter, cuts all the ordinates to that diameter in the same ratio of the segments.

For example, if the lines *KX*, *GF*, *DB*, *NA*, and *SY* are ordinates applied to

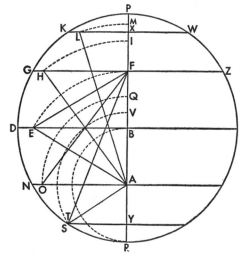

PR, and if the curved line *PLHEOTR* is an ellipse, then necessarily *DB* : *BE* : : *GF* : *FH* : : *KX* : *XL* : : *NA* : *AO* : : *SY* : *YT*.

2. The ellipse has two points, from which it is described as from centres; I am accustomed to call these two points the "foci." Accordingly if the lines drawn from the two foci to any point on the ellipse, or even the lines drawn from one focus to the points opposite the centre of the ellipse, are added together, they are always equal to the longer diameter. Hence when they are drawn to those points on the ellipse which are in the shorter diameter lying midway between the vertices, each of them is equal to the semidiameter of the circle.

For example, if *A* is the focus, *B* the centre of the circle, *AB* and *BF* equal, *F* will be the other focus. And the sum of *AH* and *HF* will be equal to the diameter *PR*. So will the sum of *AL* and *LF*, and the sum of *AO* and *OF*. Wherefore, since *BE* is the shorter semidiameter, and *E* is the point on it, *AE* and *EF* will be equal, and each of them will be equal to the semidiameter *BP* or *BR*, or *BD*.

This is applied to the planets, as follows: we have said that observations bear witness that the planets are at a distance of the semidiameter of the eccentric circle from the sun—one focus of this ellipse—at a time when they have traversed exactly a quadrant of the orbit from apsis *P*.

Demonstrate that these properties of an ellipse are exhibited in the planetary orbit which arises from these librations.

Then, in accordance with the laws so far given, let there be described a new figure, namely, with centre B, the circle PDR, to which the ellipse should be tangent. Let PR be the longer diameter of the ellipse, and on PR let A be a focus, or the place of the sun. Now let DT be drawn through B perpendicular to PR; the shorter diameter will be on DT. And because BA the eccentricity is half of the libration, therefore half the libration has been completed at the completion of the quadrant. Therefore the planet falling upon line DB will be less distant from the sun [660] than at P, and the difference will be equal to BA. Therefore it will have a distance equal to the magnitude PB. Wherefore let an interval equal to BP be extended from A to DB, and let its terminus be E. Therefore the orbit of the planet will cut DB at E.

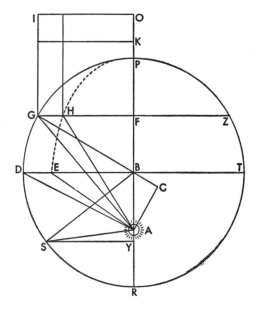

Again let there be taken PG an arc of the circle, and GFZ its sine or ordinate, and PF the versed sine. Accordingly make BP be to PF as BA half of the libration is to the part [of the libration] belonging to PG. And when that [part of the libration] has been subtracted from AP, let the remainder be extended from A to GF, and let its terminus fall upon H. I say that $DB : BE :: GF : FH$. For let square $GIOF$ be described on GF, but square HK on HF, so that the gnomon HIK is made. Then let GA and GB be joined; [661] and let the perpendicular AC be drawn to GB continued.

I say in the first place that the square on AC is equal to the gnomon HIK. For

$$BP : PF :: BA : AP-AH.$$

Wherefore too

$$PB : BF :: BA : AH-BP.$$

But too

$$PB : BF :: GB : BF :: AB : BC$$

because the right triangles GFB and ACB have their vertical angles GBF and ABC equal. Therefore

$$BC = AH-BP.$$

But too

$$CG-BP = CG-BG = BC.$$

Wherefore
$$GC = HA.$$
But
$$\text{sq. } GC + \text{sq. } AC = \text{sq. } GA.$$
But on the other hand
$$\text{sq. } AF + \text{sq. } FG = \text{sq. } GA.$$
Therefore
$$\text{sq. } GF + \text{sq. } FA = \text{sq. } GC + \text{sq. } CA.$$
Therefore
$$(\text{sq. } GC + \text{sq. } CA) - \text{sq. } GC = \text{sq. } CA.$$
But
$$\text{sq. } AH = \text{sq. } GC$$
and
$$\text{sq. } AH = \text{sq. } AF + \text{sq. } FH.$$
Therefore
$$(\text{sq. } GF + \text{sq. } FA) - \text{sq. } AH = \text{gnom. } HIK.$$

From this the remainder of the proposed demonstration is easily woven together.

For as one sine GF is to its perpendicular AC, so all the other sines are to their own perpendiculars from A. Therefore as GO the square on the sine is to the square on AC, *i.e.*, to the gnomon HIK, so the squares of all the sines are to their gnomons. Wherefore also, if the gnomons are subtracted, as GO the square of one sine GF is to HK the square on FH—which has been determined by HA the distance of the planet from the sun—so the square on any sine is to the square of its minor determined by its distance. But if the squares are proportional to one another, the sides themselves are proportional to one another. Therefore as GF is to FH, the portion terminated by AH; so any sine, such as DB, is to BE, the portion terminated by AE. And this proportionality [*ratio*] is true of the ellipse.

The other property of the ellipse is clear of itself.

For in accordance with the rule of the laws of libration, that is, because half of the libration, equal to BA, should be completed in PE [662] a quadrant of the orbit, we extend AE—which is equal to BP the remainder [of the libration] —from A to DB. For, because A is one focus, if the line extended from B along BP is set down equal to BA, the other focus will be marked out: its distance from E will be equal to AE, and the sum of the two focal distances will be equal to the diameter—and that is the case in an ellipse.

What is the ratio of DE, *the width of the crescent cut off by the ellipse from the circle, to the eccentricity* BA?

The eccentricity BA is a mean proportional between DE and ET. In the same way too, every perpendicular, such as AC, is a mean proportional between GH and HZ, the remainder of the chord.

For the rectangle GH, HZ is equal to the gnomon HIK. But this gnomon is equal to the square on AC. Therefore too the rectangle GH, HZ is equal to

138   **KEPLER**

the same square on *AC*. Therefore *GH*, *AC*, and *HZ* are in
continued proportion.

*What shall I hold concerning the
length of this elliptical orbit and
of its parts?*

If the figures of the circle and
the ellipse are cut by an infinite
number of ordinates, *GF*, *DB*
etc., the first portions ending in
P will be—*GP* will be to *PH*—
as *GF* is to *FH*; but the last portions ending in *D* and *E*—*GD*
will be to *HE*—will be equal to
one another: so the ratio of *DB*
to *BE*, which started at *P*, is
gradually obliterated, and at *D*
and *E* vanishes into the mere
ratio of equality. But the total
arcs, which started at *P*, have to
one another a ratio compounded
of all the ratios of all the least

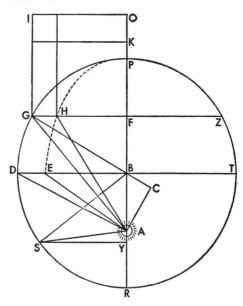

particles; and so they never lay aside completely the full ratio *DB* to *BE*. For
the quadrant *DP* is to the quadrant *PE*, and also the total circular line is to
the total elliptical line, as *DB* is to the arithmetic mean between *DB* and *BE*,
which is slightly longer than the mean proportional.

*Because use will also be made of the plane of the ellipse, I ask in what ratio will
the plane of the ellipse be to the plane of the circle; and hence the plane of any segment of the semicircle to the plane of the segment—of the semiellipse—made by the
same ordinate?*

Apollonius demonstrates in his *Conics* that the ratio of the longer diameter
to the shorter holds everywhere. Hence if *DB* and *GF* are ordinates, as *DB* is to
BE, so is the area of the semicircle *PDR* to the area of the semiellipse *PER*; and
as *GF* is to *FH*, *i.e.*, as *DB* is to *BE*, so *GPF* a segment of the semicircle is to
HPF a segment of the semiellipse; and so too *GRF* a greater segment of the
semicircle is to *HRF* a greater segment of the semiellipse.

[664] Now let the semicircle be cut by the straight line *GA*, but the semiellipse
by the straight line *HA*; there will be the triangles *HAF* and *GAF* having the
same altitude *FA*. Wherefore

> base *GF* : base *FH* :: area *GAF* : area *FAH*.

But

> *GF* : *FH* :: area *GPF* : area *FPH*.

Wherefore

> *GF* : *FH* :: *DB* : *BE* :: comp. area *PGA* : comp. area *PHA*.

*Finally I should like to know what ratio the lines from the centre of the figure to
the circumference of the ellipse have to the semidiameter of the circle?*

A very small ratio indeed, as *BE* is less than the semidiameter *BD* by *DE*
the total width of the crescent. But all the remaining lines, such as *BH*, have

a lesser difference between themselves and the semidiameter BG than at any given place, the width of the crescent, such as GH, is.

For in the triangle GHB, the sum of the two sides GH and BH must exceed the third side GB. Therefore the ratio of the defect at E to the defect at H is greater than the ratio of DE to GH. But this last is the ratio of the sine DB to sine GF. Therefore the ratio of the defect at E to the defect at H is greater than that of sine DB to sine GF.

Conversely, the ratio of the squares on GF and HF is as GF^2 is to HF^2. But

$$\text{sq. } GF + \text{sq. } BF : \text{sq. } HF + \text{sq. } BF < GF^2 : FH^2.$$

Wherefore

$$\text{side } GB : \text{side } BH < GF : FH.$$

Therefore the greater BF is, the more the ratio $GB : BH$ is diminished, so as not to equal the ratio $GF : FH$. And conversely, the more PF increases, the more the ratio $GB : BH$ increases, approaching the ratio $GF : FH$. But PF increases slowly from P, but quickly near DB. Therefore if GH everywhere stayed of the same magnitude, it would vary the defect HB slowly around P, quickly around D. But GH does not remain the same, but increases quickly around P, slowly around D—that is, it increases with its sines GF and DB. Again therefore the defect HB increases quickly around P, but slowly around E. Therefore

$$\text{defect } EB : \text{defect } HB < \text{sag. } PB : \text{sag. } PF.$$

But

$$\text{arc } DP : \text{arc } PG > \sin DB : \sin GF$$
$$< \text{sag. } BP : \text{sag. } FP.$$

Therefore the ratio of the defect, of lines BH, [665] approaches the ratio of the degrees PG. However, towards D it verges upon the ratio of sine DB to sine GF. But towards P it verges upon the ratio of sagitta BP to sagitta FP.

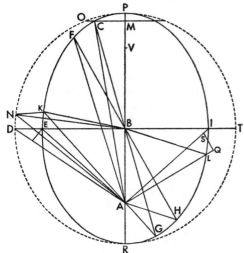

4. On the Measure of Time, or of the Delay made by the Planet in any Arc of its Orbit

How is the plane of the segment of an ellipse fitted for measuring the delay of the planet in the arc of that segment?

Not otherwise than if by the division of the circle into equal parts unequal arcs of the ellipse are constituted—small arcs around the apsides, and greater around the mean longitudes—as follows:

With centre B and distance BP, let the circle $PDRT$ be described. Let PBR be its diameter, and on *PBR*, as on the line of the apsides, let *A* be the sun, [666] the source of movement, towards *R*; *AB* the eccentricity, and *BV* equal to *AB*, towards *P*; so that *P* and *R* are the apsides.

Now with the points *A* and *V* as the foci, let the ellipse *PERI* be described, which is tangent to the circle at *P* and *R*, and which represents the orbit of the planet. And let *EI* be the shorter diameter, and *DT* the diameter of the circle, which are at right angles to *PR*.

Now let the semicircle *PDR* be divided into small equal parts, and let *P*, *O*, *N*, *D*, *R*, and *T* be the points between the divisions; and from these points let there be drawn to *PR* the line of apsides the perpendiculars, such as *OM* and *NK*, cutting the ellipse at points *C* and *K*. Therefore if the points *C*, *K*, *E*, and *I* of the sections are joined with *A* the sun, I say that the delay of the planet in arc *PC* is measured by the area *PCA*, the delay in the arc *PCK* is measured by the area *PCKA*, the delay in *PE* is measured by the area *PEA*, and finally the delay in *PER*, half of the orbit from apsis *P* to apsis *R*, is measured by the area *PERP*, which is half of the total area of the ellipse *PERIP*.

[667] *Show in what ratio, by means of this sectioning, the mean parts of the planetary orbit are made greater than the parts around the apsides.*

In the ratio of the longer semidiameter to the shorter.

For in the circle let there be the equal parts *PO* and *ND*—*PO* at the apsis *P*, and *ND* at the mean longitude *D*. Therefore, since on the sectioned ellipse the arcs *PC* and *KE* correspond to them, it has already been said above that *KE* is equal to *ND*—the most minute division being supposed—then *KE* will also be equal to *PO*. Moreover, it was said that as *OM* is to *MC*, *i.e.*, as *DB* is to *BE*, or the longer semidiameter *PB* to the shorter semidiameter *BE*, so *PO* the arc of the circle is to *PC* the arc of the ellipse; therefore as *PB* is to *BE*, so also will *KE* the arc of the ellipse at the mean longitude be to *PC* the arc at the apsis.

What follows from this sectioning of the elliptic orbit into unequal arcs?

It follows that equal areas are assigned to the sum of the [two] smaller orbital arcs taken around both apsides and to the sum of the [two] greater arcs taken around both mean longitudes, as the measures of the delays in those arcs—though nevertheless the sum of the [two] smaller arcs is as equally distant from the sun as the sum of the [two] greater arcs.

For as above let *PC* and *RG* be equal. The areas *PCB* and *RGB* will also be equal. Again let *KE* and *LI* be equal to one another, but greater than the former arcs, as has already been demonstrated: the areas *KEB* and *LIB* will also be equal.

But it has already been demonstrated that

$$PB : BE :: KE : PC$$

in the given sectioning of the orbit. Accordingly there are the triangles *BPC* and *BEK*—rectilinear or as it were reciprocal, because as *BP* the altitude of the first triangle is to *BE* the altitude of the second, so is *KE* the base of the second triangle to *PC* the base of the first triangle. Wherefore

$$\text{area } BEK = \text{area } BPC.$$

Therefore

$$\text{area } BEK + \text{area } BIL = \text{area } BPC + \text{area } BRG.$$

But

$$\text{area } BPC + \text{area } BRG = \text{area } APC + \text{area } ARG,$$

because

$$[668] \text{ alt. } BP + \text{alt. } PR = \text{alt. } AP + \text{alt. } AR.$$

And

$$\text{area } BEK + \text{area } BIL = \text{area } AEK + \text{area } AIL,$$

because on base EK or on the line touching it at E, the triangles BEK and AEK have the same altitude BE and the same base EK, and on base IL or on the line touching it at I the triangles BIL and AIL have the same altitude BI and the same base IL. Therefore the areas EAK and IAL are assigned to the long arcs KE and LI; and areas APC and ARG, which are equal to them [in sum], are assigned to the shorter arcs PC and RG, since nevertheless the sum of the distances EA and AI from the sun is equal to the sum of the distances PA and AR, as was demonstrated before.

If equal areas are assigned to unequal arcs equally distant from the sun, while the times or delays of unequal arcs equally distant from the sun ought also to be unequal—by the axiom employed above—then how do equal areas measure unequal delays?

Although in this way the pairs of arcs are really unequal to one another, nevertheless they are equivalent [*aequipollent*] to equal arcs in partaking of periodic time.

For it has been said in the above that, if the orbit of the planet is divided into the most minute equal parts, the delays of the planet in them increase in the ratio of the distances between them and the sun. But this is to be understood not of all equal parts as such, but principally of those which are opposite the sun in a straight line, as PC and RG, where there are the right angles APC and ARG. But in the case of the other parts which face the sun obliquely, this is to be understood only of that which in any of those parts belongs to the movement around the sun. For since the orbit of the planet is eccentric, therefore in order to form it two elements of movement are mingled together—as has been demonstrated already: one element comes from the revolution around the sun by reason of one solar virtue; the other comes from the libration towards the sun by reason of another solar virtue distinct from the first. For example, in arc IL the termini I and L are at unequal distances from A the source of movement; therefore let AL continued to Q as AQ be a mean proportional between AL and AI. And if with centre A and distance [669] AQ the arc QS is described cutting the longer line AI in S, arc QS is of the first element of the composite movement; but the difference between AL and AI, or the sum of LQ and SI, is of the second element of the movement, which must be separated out in mind. For none of the periodic time is due to it, since it [the second element] has already in the above received its own portion under other laws, when it was a question of the libration. But this second element of movement can be separated out in no other way than by this sectioning of the orbit into unequal parts, which we gave above. For the quantity whereby the sum of the arcs KE and LI exceeds the sum of the arcs PC and RG comes wholly from the second ele-

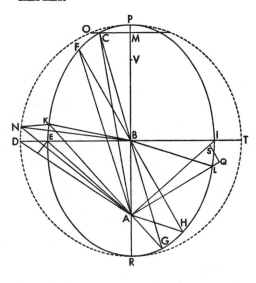

ment of the movement; and with that excess separated out, there remains of the first element something which is equal to the sum of the arcs *PC* and *RG*, which I demonstrate as follows.

For by the above demonstrations, *AE* and *AI* are equal to *BP* and *BR*. Let arcs [of circles] be described through points *E* and *I*, and let the first arc cut away and subtract as much from the area *AEK* towards *K*, as the second arc adds to area *AIL* above *L*: Hence in this fashion the triangles—or rather sectors—get these new right bases instead of the oblique bases *KE* and *LI*: Wherefore if the area equal to the sum of *PCB* and *RGB* is applied to *AE* and *AI*, the bases or arcs described through *E* and *I* are equal to the bases described through *P* and *R*. But it was previously demonstrated that

$$\text{area } KEA + \text{area } LIA = \text{area } PCB + \text{area } RGB.$$

Therefore the sum of the oblique bases *KE* and *LI* which has to do with the revolution around the sun is equal to the sum of the arcs *PC* and *RG*, where almost no libration with respect to the sun is mixed in with it, because *AP* and *AC* are imperceptibly different, like *AR* and *AG*.

The same things will also be demonstrated of the other small parts of the orbit: for example, if *CF* is taken and *CB* and *FB* are continued to *G* and *H*, and if *GH*, corresponding to *CF*, is joined, and the four points are connected with *A* the source of movement. For it was demonstrated in the above that

$$CA + AG = PA + AR = PR$$

and

$$FA + AH = PA + AR = PR.$$

Wherefore too, as previously,

$$\text{area } ACF + \text{area } AGH = \text{area } BCF + \text{area } BGH,$$

and hence

$$\text{area } ACF + \text{area } AGH = \text{area } APC + \text{area } ARG,$$

although by the proportionality of sectioning set up, *CF* [670] will be slightly longer than *PC*, and *GH* longer than *RG*. For the sum of the new arcs described with centre *A* and distances *AC* and *AG* and cutting *AF* and *AH* will be equal to the arcs *PC* and *RG*; because the greater the circle of the first arc than the circle of the second arc, the smaller an angle *CAF* does the first arc measure, and the greater an angle *GAH* does the second arc measure, so that thus

$$\text{angle } CAF + \text{angle } GAH = \text{angle } PAC + \text{angle } RAG.$$

Therefore, since the regularity of the first element in the planetary movement, namely the regularity of the progression around the sun, consists in the equality of the angles around the sun, *i.e.*, in the equality of the sum of two angles; and since the area of the ellipse is equally distributed among the arcs which subtend these angles, that is, two areas are always equal to two other areas: therefore rightly—up to now and in so far as we are dealing with pairs of arcs—is the area set up as the measure of time; because the delays of time ought to be equal not to equal arcs as such but to their equal progressions around the sun, at the same distance from the sun.

Then in this way let the area of the ellipse be rightly distributed between the pairs of opposite arcs; now demonstrate that the single triangles are separately the most exact measures of the single delays?

The demonstration is easy by means of the foregoing.

For since by our axiom the delay of the planet in arc PC is to the delay in the equal arc RG, as AP the distance of arc PC from the source of motion is to AR the distance of arc RG; but since also the area of triangle PCA is to the area of triangle RGA—which has its base RG equal to PC the base of the first triangle—as PA the altitude of triangle PCA is to RA the altitude of triangle RGA: Wherefore the delay of the planet in arc PC is to the delay in the equal arc RG as the area of triangle PCA is to the area of triangle RGA.

In the same way it will be demonstrated that the delay of the planet in CF —which is equal in power (*potestate*) to CP—is to the delay of the same in GH, as the area ACF is to the area AGH—where the sum of each pair of areas is equal to the sum of the [two] prior areas, and so on in order. Therefore the total area of the ellipse—which has been cut up at A into triangles—[671] is distributed among the arcs in the same proportion wherein the total periodic time has been distributed among them. Therefore the single triangles are proportionally the most exact measures of their single arcs.

A demonstration of this full equivalence is given in my *Commentaries on Mars*, Chapter 59, page 291. On that page at the line *Apsis longiorem*, one word, *erit*, has brought in great obscurity; and if you change it to *computaretur*, everything will be clearer. Although I confess that the thing is given rather obscurely there, and most of the trouble comes from the fact that there the distances are not considered as triangles, but as numbers and lines.

5. On the Equivalence of the Circular Plane and the Plane of the Ellipse in Measuring the Delays in the Arcs

Does it not seem to be a hard, unaccustomed, not to say complicated, job for the calculator to be reduced to using the plane of the ellipse in computing the time?

On the contrary, in the opinion of all the job becomes easier by the employment of the elliptical plane instead of the circular; so much so that the ancient calculus is by no means to be compared to this new one for ease.

Demonstrate the equivalence of the planes with respect to measuring time.

Then let there be repeated the figure exhibited on page 138 where we demonstrated the generation of the plane of the ellipse.

And because it has already been demonstrated that as half of the periodic time—wherein the planet traverses *PER* half of the orbit—is to the time which the planet spends in *PH* or in *PE*, precisely so is the area *PER* to the area *PHA* or *PEA*. But it was also demonstrated above that

PDR : *PER* :: *PGA* : *PHA* :: *PDA* : *PEA*.

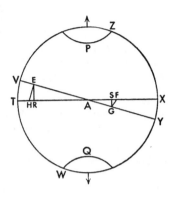

For they all have the same ratio as [672] *DB* to *BE*. And thus too, by alternation, as the area *PER* is to *PHA* or to *PEA*, so the area *PDR* is to *PGA* or to *PDA*. Therefore as half the periodic time of arc *PER* is to the time of arc *PH* or *PE*, so the area *PDR* is to *PGA* or *PDA*. Wherefore in these segments of the semicircular plane there is found a very exact measure of the delays which the planet makes in any given arc of the ellipse.

Now also show the convenience of this measuring.

Let the segment *PGA* be taken, and let a straight line be drawn from *G* to the centre *B*. Therefore the ratio of the sector *GBP* to the whole plane of the circle is given by the given magnitude of the arc *PG*—so that there is no need of computation. For the total periodic time and the total plane of the circle are divided into 360°—astronomical fashion. Therefore *GBA* the other part of the segment is left. But the computation of it is easy. For as *DB* the total sine is to *GF* the sine of the given arc *PG*, so is area *DBA* to area *GBA*. Therefore once the area of the great triangle *DBA* has been determined, that is, by the multiplication of half the eccentricity by the total sine and by the conversion of the result into astronomical terms, it will ever afterwards be useful.

Does not the circular plane have some other further use?

In the schema of the moon there is a special use of it in demonstrating one of its irregularities, which the moon possesses uniquely, in contrast to all the other planets. But because this Fifth Book is given over only to those properties which are common to all the planets; therefore what remains of the geometrical apparatus for carrying out the demonstration of this special use is rightly postponed to Book VI Part IV, *i.e.*, to the schema of the moon.

In what way does the old Ptolemaic astronomy measure the delays of the planet in any arc of its eccentric circle, or what does it have instead of the circular plane?

For this it uses a special circle, which has the name of "equant," [673] of which the centre would in our figures be the other focus—in the last diagram, *F*, in the next to the last, *V*; because the other focus is as far distant, towards *P* the highest apsis, from *B* the centre of the eccentric circle, as *A* the sun, towards *R* the lowest apsis, is distant from the same centre of the eccentric circle. For if a line is drawn from *V* the centre of the equant through the body of the planet, the arc of the equant intercepted between this line and *VP* the line of the apsides is set down as the measure of the time which the planet spends in the arc of its orbit.

This hypothesis seems to be more convenient for manual representations by means of the instruments called schemata: why do you not keep it, since you have already twice employed substitute quantities in place of the true?

1. Because the equant never says the truth perfectly, unless we wish to give its centre an irregular movement of libration. And in that way we should draw away from simplicity of hypotheses, and set up an astronomy which is more complicated and laborious in practice than the astronomy of these two books, the fourth and fifth, in the explanation of causes. And with these causes once perceived, and not even believed in but merely laid down, practice afterwards becomes easy in the second part of Book v and in Book vi.

2. Because in Ptolemy the proportionality (*ratio*) of this equant is one thing in the higher planets, something else in the two lower planets, something else in the moon, and now it would be something else again in the sun. But for us the plane of the eccentric circle serves the same use in the same way in all the planets.

3. Because the equant circle is very distant from the true causes of the movements. And the plane of the [eccentric] circle represents these causes very closely, because it is of the same family as the plane of the ellipse.

Let it be understood that the same things are spoken against other equivalences, which the wonderful force of human ingenuity is accustomed to bring forward: for example, by means of a single libration of the centre of the eccentric circle in the shorter diameter of our ellipse—although the libration needs the contrary motions of two equal circles—[674] David Fabricius both saves the movements inward of the planet away from the sides (*a lateribus*) of our immovable eccentric circle and at the same time gives the apsis a movement of libration—so that now, by means of numbering continued from the apsis, which has the libration, to the planetary body, the eccentric circle itself furnishes us with a measure of time. For neither absolute regularity of movements nor perfect precision is obtained, nor is there a curtailment of labour. And the causes of the movements are hidden and denied.

But I absolutely reject the Copernican machinery, which makes two epicycles having the ratio of one to two in their movements revolve on a concentric circle. For observations bear witness that at the mean positions between the apsides the planet moves inward towards the sides. But this Copernican hypothesis by a contrary proportionality [*ratione*] makes it wander outward.[1] These small parts of the Copernican hypotheses must be completely corrected. But his general hypothesis of the annual movement of the Earth must be saved, whence the name of this teaching comes.

6. ON THE REGULARITY OF THE DIGRESSIONS IN LATITUDE

The calculus of latitude is not certain, is it, if there are no solid spheres, even though special threads in the planetary body take their place?

If we lay down those things which were laid down in Book IV, page 108, and

[1] See *Revolutions of the Heavenly Spheres*, Book v, Chapter iv.

which are absolutely possible and concordant, then it is absolutely necessary for the perfect plane of an ellipse to arise.

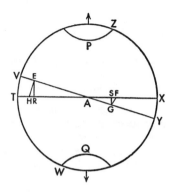

For in this diagram let *TZX* be a circle through the poles of the ellipse. Either let *A* be the sun, if *TZX* is a plane; or, if *TAX* is a hemisphere, let *A* first be the position, on the concave surface, of the lower intersection of the ellipse *TX* and *EG* the orbit of the planet, —so that the poles of the orbit are at *Z* and *W*. Let the threads of latitude point along *GA* [*dirigantur secundum* GA]; and let them have the faculty of deflecting the movement *XAT*, given by the sun, through the angle *GAX;* and during the whole circuit let the threads remain parallel. [675] It is clear that while the planet is turning at *A* the lower section, the threads stretched along *GA* will direct the planet according to the total angle, and the planet will come in a perfect plane as far as *G*, by ascending up to the plane drawn through the poles. And because now the thread at *G* points towards the sun *A* and not crosswise over the ecliptic; accordingly here the planet will not digress any farther, but *G* will be the limit: thence, gradually raised up above the plane *ZXW*, the planet will direct its thread along the line drawn from intersection *A* through the sun *A*, until the planet comes to *A*, now the upper intersection on the convex surface. Therefore just as at *A* there is the greatest angle of inclination of the thread with the ecliptic *TX*, and the angle decreases rapidly; but at *G* and *E* there is no angle of inclination of the thread with the plane [*longitudinem*] of the ecliptic, and this smallness of inclination lasts a long time; so too if out of the circuit *EAG* there is made a perfect plane, its parts at *A* have their greatest inclination with the ecliptic *TX*, and the inclination decreases quickly. But around *G* and *E* the rim of the plane —which is understood to stretch downward into the depth of the sphere or upward—extends for a long time approximately parallel to the plane of the ecliptic. Therefore if instead of the operation of the thread we employ the product itself, that is, *EAG*, as a perfect plane, the calculus will be absolutely consonant with the principles.

Conclusion of the First Part of Book V

Then let what has been written so far be done for geometers endowed with a keen mind, who do not think it right to admit into their calculus that which has not been confirmed by a very accurate demonstration and deduced from the natural principles of the movements.

Part Two

On the Astronomical Terms Arising from Calculations and the Eccentric Orbit

[676] *How is the orbit of any planet called?*

It is called by the ancient name of "eccentric," *sc.* circle. For even if the orbits

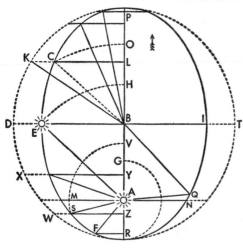

are elliptical, as *PERI* here, and have as it were two centres, *A* and *L*, which, in physical language, we call the "hearths" or "foci," and the sun itself, as centre of the world, is situated at one of those foci; nevertheless also the point midway between the foci, as *B*, is with peculiar [677] rightness called the "centre" by writers on conics. And furthermore, for the sake of measurement, a perfect circle *PDR* having its centre *B* distinct from *A* the centre of the world is circumscribed around the figure.

In astronomy, what name does PR *the longer diameter of the ellipse have?*

It is called the "line of apsides," because, since it is drawn through *A* the centre of the world and *B* the centre of the orbit, its sections with the orbit designate *P* the highest apsis and *R* the lowest.

Why are they called the highest apsis and the lowest, and what other name do they have?

The word "apsis" is taken from wheels; for the apsides are points on the eccentric, *P* the farthest away from *A* the sun and *R* the nearest to it. But in geometry the explanation of the meaning is clearer. For the word "apsis" is derived from "being tangent"; and indeed in points *P* and *R* the director circle is tangent to the elliptic orbit. The Latin translations of Arabian books express the Greek words "apsis," "apsides" by the words "aux," "auges"—as if the Arabs had changed the Greek Psi to Xi. However some one who boasts a knowledge of the Arabic language asserted to me that the word "augh" signifies altitude.

Book VI will call those points the "aphelion" and the "perihelion," in the case of the primary planets; and in the case of the moon, the "apogee" and the "perigee."

What necessity compels us to suppose, instead of the circular route of the planet believed in by the ancients, an elliptical route, i.e., one falling away from the circle, and to set up in it a longer diameter and on that longer diameter the sun?

Both of them are demonstrated from the observations and by means of a very sure demonstration in the *Commentaries on the Movements of the Planet Mars*, and they are employed in Book IV on page 22 in the diagrams and on pages 70 and 92 and 93, and also in the first part of Book V. Therefore, unless we made these hypotheses, we should never represent the observations.

[678] *By what names do we distinguish from one another the semicircles into which the line of apsides divides the eccentric?*

The one half, *PER* or *PDR*, is called the descending or the prior semicircle; the other, *RIP* or *RTP*, the ascending or posterior.

What is eccentricity?

In Greek, ἐκκεντρότης is the line joining *A* the centre of the world, or of the body around which the movement is arranged, and *B* the centre of the eccentric—namely *AB*, part of the line of apsides *PR*.

What is the name of the lines drawn to the eccentric orbit from the centre of the body around which the movement is arranged?

In Greek, they are called ἀποστήματα; as latinized, "intervals" or "distances"; in translations from the Arabic, "elongations"—for example, *AP*, *AC*, *AE*, *AM*, *AS*, *AF*, *AR*, *AN*, *AQ*, etc.

[679] *Which among these elongations are the more important?*

AP, the longer elongation, in Arabic, or the aphelial or apogeal distance; *AR*, the shorter elongation or perihelial distance or perigeal distance, in the case of the moon; and the mean elongation, which is the arithmetic mean between the longer and shorter elongation: the one which is in the descending semicircle, namely *AE*, is called the first mean elongation, and the one in the ascending semicircle, such as *AI*, the second.

What else does "mean elongation" signify?

By metonymy, it is used for those points on the orbit which are at a mean distance from the sun, such as *E* and *I*, namely those points which are at a distance of a quadrant or 90° before or after the apsides. And sometimes it is also used for the point in the zodiac which is 90° distant before or after the position in the zodiac on which the extension of the line of apsides falls.

Here we must note emphatically that it is not the extension of line *AE* of the same name which falls/upon this degree of the ecliptic called the mean elongation, but rather line *BE* from the centre, or its parallel, *AM*—as being lines which make right angles with *PR* the lines of apsides.

What is the name for the difference between the mean elongation or distance and any other whatsoever?

This difference is called the "libration" of the planet, because the total libration, as in the movement of the trays of the balance, is slow in the beginning, when the planet is at its greatest distance from the sun, and in the end, when the planet is nearest the sun; it is fast in the middle.

In the diagram, since *AP* is the longest distance, and *AR* the shortest, then let *AR* be transposed to line *AP* and extended from *A* to *G*, so that the total libration can be laid before the eyes in a single line *AP* which is as it were at rest; the total libration will be *PG*, twice the eccentricity *BA*. Accordingly this libration is slow around *P* and *G*, [680] namely when the planet is either in *P* or in *R*, fast around *H*, when the planet or line *AH* has been transferred to line *AE* or *AI*.

You have said that a circle is circumscribed around the orbit for the sake of measurement. State in how many ways this circle conduces towards the measurement of that orbit.

Four ways: (1) This circle designates and divides the arcs of the elliptic orbit. For example, arc *PC* receives its name and specification from arc *PK*.

(2) The circle gives the measures of the librations of the planet and thus determines the lengths of the intervals between the planet and the sun. For example, *AC* or *AO* is determined by arc *PK* or its complement *KD*, because arc *PK* gives the magnitude of the libration *HO* which must be added to the semidiameter *AH*.

(3) The circle exhibits also the measure of the time which the planet spends in any arc of its elliptic orbit. For example, we learn through arc *PK* how long the planet moves in arc *PC*.

(4) When those things have been found, we can also investigate the angle at the sun subtended by the arc of the orbit. For example, if arc *PK* is not known, if line *AC* is unknown, angle *CAP* cannot be found.

1. Concerning Designation

In what way does the circle designate and divide the arcs of the ellipse, and by what means, and wherefore?

Since the circumference of an ellipse cannot be divided geometrically into equal parts, nor can parts into which it has been divided be designated from their number: therefore, in place of the ellipse, the circle is divided into equal parts—starting from the apsides, and from the points of division perpendiculars cutting the ellipse are drawn to the line of apsides. Therefore the arc of the circle between the aphelion and any perpendicular whatsoever supplies the name for the arc of the ellipse intercepted between the same limits and applies its number of degrees and minutes to that arc.

[681] Let arc *PK* be 50°0′, and *KL* cutting the ellipse in *C* perpendicular to *PR*. Therefore too *PC* the arc of the ellipse is said to be 50°0′.

But the name is untrue, since the arc of the ellipse is not so great—neither with reference to the circle nor with reference to the total elliptic orbit.

This causes no trouble, for there is no question of anything at present except the name; and the name is not the name of the apparent measure, but of the geometrical determination and division. And there is no need to know the true length of the arc of the ellipse as if measured with a yardstick, provided we may afterwards learn how great an angle is made at the centre of the sun by the arc of the ellipse so determined and how long the planet delays in that arc. As a matter of fact, in the first part of Book v, I demonstrate that this arc of the ellipse is as great, if not in length, at least in power.

How are these perpendiculars called which cut the ellipse?

In the circle, they are called the sines of the arcs of the circle which begin at the aphelion; in the ellipse, they are generically called ordinates, *viz.*, to the axis. For example, here *KL* is the sine of arc *KP*. *CL* is the ordinate.

But specifically, that which is drawn through the centre of the figure—as *EBI*—is called the shorter diameter or minor axis of the figure. We can employ the Greek term "diacentre." Finally, the perpendicular, such as *MAN*, which passes through the centre of the sun is without a name, although it is among the principal ones. Let it be called by the new name "dihelion."

Now what is the function of those two perpendiculars, the diacentre and the dihelion?

They divide the orbit into upper and lower parts: the diacentre, into parts which are equal but unequal in respect to time and apparent movement; the dihelion, into parts which are [682] unequal both in time and in length but which none the less appear equal as it were from the sun.

For example, *EPI* which is constituted by *EBI* is 180°, but it appears to be less than 180° by angle *EAI*. But the greater segment *MPN* cut off by line *MAN* and *MRN* the lesser segment both appear equal in magnitude to 180°.

2. Concerning Libration

Tell how to measure and compute the librations and how to determine the intervals.

Let *PK* be an arc of the eccentric less than 90°, for example, 46°18'51"; therefore *KD* its complement will be 43°41'9", and *BL* its sine, 69.070; and let eccentricity *AB* or half libration *PH* be 9.265, whereof *BP* is 100.000. Therefore

$$9.265 \times 69.070 = 6.399,$$

if the last five decimals are ignored; and hence the libration *OH* will be 6.399, which is to be added to *BP* or *AH* in the upper semicircle; and *AO* or its equal *AC*, namely the distance of the planet from the sun will be 106.399 (whereof the semidiameter is 100.000) which goes with arc *PK* or *PC*.

But if the arc of the eccentric is 313°41'9", the excess over and above three quadrants or 270° will also be 43°41'9", which gives the same sine as a multiplicand. Hence when the libration of 6.399 is produced, that same number is to be added to the upper semicircle which is an ascending semicircle.

But if the semidiameter *BP* receives a different mensuration, for example, 152.342, we shall multiply *BP* by the 1.06399 of *AC*, and (if the last five decimal places are ignored) *AC* will be 162.090 at this distance.

By means of Napier's invention, this whole operation is finished most expeditiously in a single addition. For the logarithm of the sine of arc *KD* is added to the logarithms of 9.265 the eccentricity and of 152.343 the mensuration proposed; and the sum produced as a logarithm will give a libration of 9.748 to be added to the distance 152.342.

[683] So next, if arc *PW* is greater than 90°, viz., 133°39'7"; *DW* its excess over and above 90° is 43°39'7"; and its sine or logarithm, together with the two aforesaid beginnings, will give a libration of 9.777 to be subtracted from 152.342, as in the semicircle below the diacentre: hence the corresponding interval *AS* will be 142.565.

The same thing will hold, if the arc of the eccentric is 226°20'53". For its complement with respect to 270° will be 43°39'7"—as great in the ascending semicircle as was *DW* in the descending semicircle.

Review the principal moments in the libration.

1. When the planet begins to move away from the apsis, then simultaneously the libration commences—that is, the planet, which had just before finished its ascent away from the sun, begins to descend towards the sun.

2. When the planet is 60° away from the apsis, then the libration is equal to half the eccentricity.

3. When the planet has traversed 90° of the orbit from the apsis, then half of the libration has been completed, so that the planet is at a distance from the sun of a semidiameter of the eccentric. For example, if *PD* is 90°, then *AE* equals *BD*.

4. When the planet has traversed 120° from the apsis, three quarters of the libration have been completed.

5. When the planet is at the lowest apsis, then it is nearest to the sun and has completed the total libration. The order through the ascending semicircle is the reverse.

6. If the planet has traversed equal arcs on the eccentric, on one occasion from the aphelion, on the other occasion from the perihelion, then the sum of those two distances from the sun is equal to the diameter. For example, if a straight line is drawn from *B* to *Q*, then the sum of *CA* and *AQ* is equal to *RP*.

3. CONCERNING THE DELAY OF THE PLANET IN ANY ARC

What does the word "anomaly" mean?

Although, properly speaking, anomaly—irregularity—is an affect of the movement of the planet, nevertheless astronomers employ this [684] word for the very movement in which the irregularity is present. And since the following three measurable things come together in the movement: the space to be traversed, the temporal delay in the space, and the apparent magnitude of the space; the word "anomaly" is to be applied to all three. And again, with reference to time, the use of the word is twofold. For, *primo*, Ptolemy employs it for the total time which the planet spends while its total irregularity returns to its starting point, and he counts as many anomalies as there are periodic returns such as these.

Secundo, parts of this total time are commonly called anomalies, according as Ptolemy said a movement was a complete part of the *full* anomaly.

Then how many anomalies are there as parts of a whole?

Three anomalies are denominated in any position of the planet whatsoever: (1) mean anomaly; (2) anomaly of the eccentric; and (3) the corrected anomaly.

What is the mean anomaly?

It is the interval of time which the planet spends in any arc of its orbit—beginning at the apsis: this interval is reduced to degrees and minutes, whereof the total anomaly is equivalent to 360° by mathematical or astronomical enumeration.

Why is it called mean?

Not from being, as it were, a mean proportional between its related anomalies, as will be remarked a little later; but it is called mean in imitation of the ancient astronomy which usually speaks of the mean anomaly instead of the mean—*i.e.*, uniform—movement of anomaly, because the time thus reduced to mathematical terms indicates, by means of the number of degrees and minutes, how great an arc of its circle the planet would have been going to traverse, if during that whole time which we call the mean anomaly it had moved in a uniform movement which was a mean between the slowest and the fastest movement.

How should the mean anomaly be determined or measured in these diagrams, according to the ancient astronomy?

If the distance *BL* equal to the eccentricity *AB* is marked off in the line of apsides *BP*, as was said in Part I of Book V; the mean anomaly, according to the ancient astronomy, would be the arc of the equant circle described around *L*—the arc which progresses eastwards and is intercepted by two lines drawn from *L*, the one through apsis *P* and the other through the planetary body *C*. Or it would be the angle between those lines at *L*, or the difference between that angle and 360°. For example, in this case if *C* were the planet, then angle *PLC* could be used instead of the mean anomaly approximately.

Define the line of mean movement and the mean position of the planet, according to this ancient hypothesis of the equant.

It would be the line drawn from the centre of the sun to the sphere of the fixed stars parallel to the line which has been drawn from the centre of the equant, or from the other focus of the ellipse, through the planetary body. And either one of these lines in the sphere of the fixed stars would indicate the mean position of the planet. In the diagram, if *C* were the planet and *AM* were parallel to *LC*, then *AM* would be the line of its mean movement.

If therefore in this new formulation of astronomy, no equant circle is expressed, then in terms of what other magnitude will the mean anomaly be computed or measured?

In terms of the area comprehended between the arc of the circle which designates and determines the proposed arc of the orbit and between the two straight lines which connect the extremities of the arc with the centre of the sun. For example, if the proposed position of the planet is *C*, then if a line be drawn from *C* perpendicular to *PR* which will cut circle *PD* in *K*, and if *PA* and *KA* are joined, area *PKA* is the measure of the mean anomaly, whereof the area of the whole circle is equivalent to 360°.

[686] *Tell how to compute the mean anomaly or the amount of time which the planet spends in the proposed arc.*

Again, let eccentricity *AB* be 9.265, whereof the semidiameter *BP* is 100.000. First of all, we must take the area of the greatest triangle which has a right angle at *B* and an altitude *BD*. The multiplication of this area by half *AB* will give a product of 4632.50000. The value of this area *DAB* is to be expressed by the number of seconds, whereof the total area of the circle *PDT* is 360° or 21,600′ or 1,296,000″. Therefore, since if *BP* is 100.000, the area of the circle by calculation is 314,159.26536, area *DAB* is 19,110″.

Now let arc *PC* be given by arc *PK* which designates it; and let it be 46° 18′51″. Therefore the sine of *PK*, namely *KL* the altitude of triangle *BKA*, multiplied by the value of the greatest triangle—and the last five decimals cut off from the product—will give triangle *AKB* a value of 3819″, which is 3°50′19″. But sector *KBP* is equal in value to the number of degrees which have been given in arc *PK*, namely 46°18′51″. Therefore by the addition of the areas, *PKA* is 50°9′10,″ and the mean anomaly is of that magnitude.

In this way, the area of the additosubtractive triangle (*trianguli aequatorii*) is to be added, as long as the sector or arc is smaller than a semicircle; but if the arc is greater than a semicircle, the area is to be subtracted.

State the rule for the relation between these triangles.

Any two triangles equally distant from the vertices, the one from the highest apsis, and the other from the lowest apsis, are of equal magnitude. For example, if arcs PK and RW are equal, areas BKA and BWA will also be equal.

What is the anomaly of the eccentric?

It is the arc of the eccentric circle measured eastwards and intercepted between the line of apsides and the perpendicular drawn to that line through the planetary or through any proposed [687] point on the orbit. For example, with C as the proposed point on the orbit or with the planet revolving at that point, if through C perpendicular to PAR line KCL is drawn cutting the circle in K, arc PK will be the anomaly of the eccentric.

In what sense is anomaly of the eccentric used?

Here too the phrase "of movement" is understood [between "anomaly" and "of the eccentric"]. For although, according to the figure, in arc PK of the circle no irregularity or anomaly is apparent: nevertheless the movement of the planet in orbit PC is truly anomalous or irregular in three respects: first, by reason of its elliptic figure which bends with unequal curvature according to the difference of its parts and is unequally distant from the centre of the figure; second, by reason of its speed, which is not the same in all parts of its orbit; thirdly, by reason of its apparent movement as seen from the sun, because equal parts of the orbit subtend unequal angles at the sun. Therefore since the arc PK concurs in all those determinations, as has been said: wherefore with the same right wherewith the ancient astronomy introduced the equant circle and by means of that computed the mean anomaly, we ourselves circumscribe an eccentric circle PK around the real orbit PC and by means of that compute the anomaly of the eccentric, employing something uniform in measuring that which is not uniform.

And indeed in the ancient astronomy the equant circle seduced the physicists into imagining that either the circle or at any rate the movement was real: but here no one can be seduced, since it is apparent to the eyes that PC the true orbit of the planet coincides with this artificial circle PK only in the two points P and R, the apsides, and throughout the remaining tract it betakes itself, within the embrace of the eccentric, towards the centre of the figure.

What is the corrected anomaly?

It is the arc of the great circle in the plane of the ecliptic which is marked out by the continuation of the plane of the planetary orbit: this arc is measured eastwards from the position of the apsis to the very position of the planet or the apparent position of any other point on the orbit. Or—which amounts to the same thing—it is the angle which any arc of the true planetary orbit subtends or which is formed by the two aforesaid lines at the centre of the sun, or the difference between that angle and 360°.

For example, if the planet is in C, the corrected anomaly is angle PAC. And if the planet is at Q, then the corrected anomaly is made up of the two right angles PAM and MAR and also angle RAQ. But if with centre A a circle of any magnitude whatsoever be described, and thus even a circle in the sphere of the fixed stars, the arc of this circle measured eastwards from AP as far as AC or continued to AG will also be called the corrected anomaly.

Why is it called "corrected" [coaequata]?

Practitioners used to call it the corrected movement of anomaly or simply the corrected anomaly, not as if the irregular movement assumed had been corrected in such a way as to become a regular [*aequalis*] movement, but with a plainly contrary meaning: Now since in the beginning a time or portion of the periodic time is laid down, and since this time, as reduced to astronomical terms, indicates how great an arc of its circle the planet would have been going to traverse within this interval of time, if it had been moving with a uniform movement: so it is the office of the astronomer to show how much of the really irregular apparent movement of the planet corresponds to this time and this fictitious uniform movement. Therefore "corrected movement" means the same thing as movement to which an additosubtraction has been applied and which has been converted to an apparent movement, namely by putting on the irregularity which the appearances introduce into it: on account of this irregularity the total period is called the anomaly.

Therefore since you have distinguished these three anomalies and formulated them by means of a fictitious eccentric circle circumscribed around the orbit: I ask whether the true orbit of the planet could not be put to the same use.

Although there is no need, nevertheless that is possible on account of their equipollence. For, as has been said in Book v, Part i, the area PCA measures the time and hence the mean anomaly; and anyone who wishes to can understand the anomaly of the eccentric circle also by arc PC. But angle PAC has even before this been called the corrected anomaly.

How are these three related anomalies distinguished in magnitude?

The number of degrees and minutes of the anomaly of the eccentric is always a mean between the others. But before a semicircle has been completed, the so-called mean anomaly is always the greatest of the three, and the corrected, the least; but after the semicircle, the so-called mean is the least in magnitude and the corrected the greatest.

4. CONCERNING THE ANGLE AT THE SUN

Tell how to compute the corrected anomaly or the angle at the sun.

There are various modes; but the most compendious is the one which [690] employs the interval between the planet and the sun. For we need that for other uses too.

Now there are three cases of this mode. For either the planet is above the diacentre, or below the dihelion, or between the diacentre and the dihelion.

1. Therefore first let the planet be above the diacentre DBT, *viz.*, in C; and let PK the anomaly of the eccentric be $47°42'20''$, and let LB the sine of its complement KD be 67.277, so that the libration of the planet will be 6.233. By the addition of 6.233 to BP let AC the interval between the planet and the sun be exactly 106.233 in terms whereof BP is 100.000. Accordingly let this same LB the sine of the complement be added to the 9.265 of eccentricity BA, in such

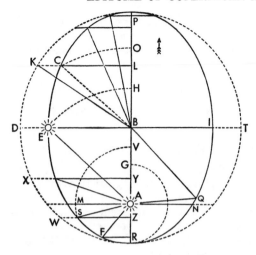

fashion that *LA* the other side of right triangle *CAL* is 76.542. Accordingly if *LA* is divided by *CA* and five ciphers are added, the quotient of 72.051 as a sine shows the arc to be 46°5'48", which is equal to angle *LCA*, whereof the complement, which is 43°54'12", is the required angle *LAC* or *PAC*.

If you subtract the logarithm of half the divisor from the logarithm of half the dividend, the logarithm of the same sine or arc remains.

2. Secondly, let the planet be below the dihelion *MAN*, viz., in *S*; and let *PW* be the anomaly of the eccentric and *DW* its excess over and above 90°. Accordingly in the same way as above, the libration found by means of *BZ* the sine of that arc is to be subtracted from the radius, so that the exact interval *AS* may be manifest. So too the eccentricity *BA* is now to be subtracted from sine *BZ*, so that *AZ* the other side of the right triangle remains. Then once more, if the number of side *AZ*—to which five ciphers have been added—is divided by side *AS*, the quotient will be the sine of angle *ASZ*, to which angle *MAS* is equal, which is the excess of the required angle *PAS* over and above angle *PAM* which is right or 90°.

3. Thirdly, let the planet be between *DBT* and *MAN*. Accordingly let *PX* be the anomaly of the eccentric and *DX* its excess over and above 90°, and *BY* the sine whereby the subtractive libration is computed, since the sine lies below *B*. But since the sine is less than eccentricity *BA*, it is to be subtracted from this eccentricity, so that *YA* remains. Accordingly we must operate on *YA* and on the exact interval, as in the first case.

[691] *What do you call the eccentric position of the planet?*

That point in the zodiac on which there falls the straight line drawn from the centre of the sun through the planetary body.

What is the "equalization" [aequatio] or "additosubtraction"[prosthaphaeresis], and what is the reason for the name?

It is the difference between the number of degrees and minutes of the mean anomaly and the degrees and minutes of the corrected anomaly. Or, according to the old astronomical formulation, it is the angle at the centre of the sun and its measure, the arc of the great circle in the sphere of the fixed stars intercepted between the lines of the mean and of the apparent movement of the planet. Now since in one semicircle this angle is to be subtracted [from the mean movement] and in the other semicircle it is to be added to the mean movement, so that the movement may be corrected: hence it has been called by the compound

name "additosubtraction" [προσθαφαίρεσις]. But it has been called the "equalization" because by its addition to, or subtraction from, the corrected anomaly,
which divides unequal arcs and times into equal portions, there comes to be
the mean and uniform anomaly.

By what epithet or title is the additosubtraction called?

By two words or their index letters or syllables: A., add., additive; S., sub.,
subtractive.

How many parts are there to the additosubtraction and what is the measure of each?

There are two parts, the one physical, and the other optical, to the aforesaid
additosubtraction. For the physical part is due to the irregularity which really
accedes to the planetary movement on account of physical causes. But the
optical part is due to an irregularity which is merely apparent or as it were apparent, *i.e.*, on account of the greater or less distance between the arc of the
orbit and the sun. Both parts are somehow distinguished in the same triangle,
which is hence called additosubtractive [*aequatorium*].

For if A and B the termini of the eccentricity are joined to [692] the planetary
body C, the physical part of the additosubtraction finds its measure in area BAC
—or by equipollence, in area BAK; while the optical part of the additosubtraction would be equal to angle BCA, if that were computed: angle BKA,
which is easier to compute, is always slightly less than angle BCA.

What use is there for this additosubtraction?

In this formulation of astronomy made new, there is no necessary or very
great need of the total additosubtraction composed of both elements. For the
anomalies are not determined through this additosubtraction; but on the contrary through the comparison of the corrected anomaly—which we first compute—with the mean anomaly, we elicit the additosubtraction, if at any time
we wish to use it.

But three distinct anomalies are laid down in the tables. For, first, the anomaly of the eccentric is placed on the left, according to the order of the whole degrees, from 1° to 180°; and that is done [693] because the anomaly of the eccentric being given becomes the starting-point for computing the others, and also
the distance or interval between the planet and the sun. Secondly, there is
found under the same head and as corresponding to this anomaly of the eccentric, the physical part of the additosubtraction or the value of the area of the
additosubtractive triangle in degrees and minutes and seconds: and from this
inclusion of the anomaly of the eccentric together with the physical part of the
additosubtraction beneath the same head, we understand that the sum of the
two constitutes the corresponding mean anomaly. Thirdly, to the right of this
and in a separate column is placed the corrected anomaly corresponding to the
arc. If anyone now wishes to learn the composite additosubtraction, let him
subtract the corrected anomaly from the mean anomaly found next to it or
from the sum of the anomaly of the eccentric and the physical part of the
additosubtraction; and the remainder will be the required additosubtraction,
which in the descending semicircle is called subtractive, and in the ascending,
additive.

Nevertheless state how these parts of the additosubtraction are related, if they are compared with one another.

The smaller the eccentricity, the nearer they approach to equality: in the upper semicircle however, above the diacentre, the optical part is slightly smaller than the physical part; but in the lower semicircle, below the diacentre, it is slightly greater.

For example, in the accompanying diagram, let *A* be the sun and *PAR* the

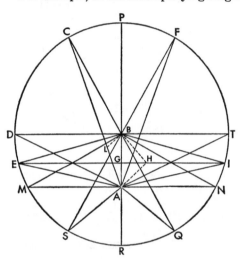

line of apsides, and *DBT* and *MAN* are at right angles to *PAR;* and *DPT* the upper semicircle, or so to speak, and *DRT* the lower semicircle. In the upper semicircle let the additosubtractive triangles be *BCA* and *BFA*, and *BSA* and *BQA* in the lower. Therefore since the areas of the triangles are the measure of the physical part of the additosubtraction, while the angles at *C*, *F*, *S*, and *Q* are the measure of the optical part: then surely the upper areas are greater parts, while the lower areas are lesser parts, of the area of 360° of the total circle than those angles of theirs are of four right angles or 360°. For the centres *C* and *S* and *CB* and *SB* as the lengths of the semidiameters, let there be described the arc *BL* ending at *CA* and the arc *BH* ending at *SA* continued: these arcs measure angles *C* and *S*. But areas *CBL* and *SBH* are equivalent to these same arcs. Therefore if these areas were the optical [694] parts of the additosubtractions, the two parts of one additosubtraction would be equal. But not *CBL* but the greater area *CBA* is the measure of the optical part; and thus, in the lower semicircle, not *SBH* but the smaller area *SBA* in the lower. Therefore in the upper semicircle the physical part exceeds, and the optical part in the lower.

Where is the composite additosubtraction greatest?

Of the parts, the prior one, the physical, is greatest at *D* and *T*, the extremities of the diacentre, because the altitude of no triangle can be greater than *BD* or *BT*, which is the semidiameter in the circle and even in the ellipse, the longest of the ordinates. The posterior part, the optical, would be greatest, if the orbit were circular, at *M* and *N* the extremities of the dihelion: for there the line drawn from centre *B* perpendicular to the straight line *MA* would be the longest, but that line is the sine of angle *BMA* the optical part, whereof *BM* is the sine of the total additosubtraction. For upon *EA*, a higher line, there falls from *B* a shorter perpendicular than *BA*.

But because the orbit of the planet is elliptical, accordingly the optical part of the additosubtraction is [695] greatest between M and D, and thus between N and T. For, first, angle BMA is greater than angle ADB, because both are right triangles upon the same base, but altitude DB is greater than altitude MA, namely the shorter diameter is greater than any other ordinate whatsoever. Second, if points B and I are marked at the centres of arcs DM and TN or thereabouts, angles AEB and AIB will again be greater than AMB and ANB. For, of all the lines drawn from centre B to the orbit, BD is the shortest; and the farther away the other lines are, the longer they

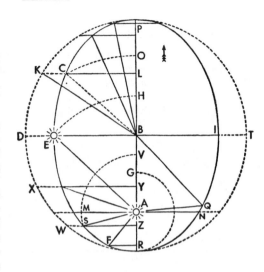

are. Therefore BM is perceptibly longer than BE. But the perpendicular from B to AM is not perceptibly longer than the one from BE to AE. Therefore the ratio of MB to BA is greater than the ratio of EB to its perpendicular. And so angle BEA is also greater than angle BMA. Therefore if BA is bisected at G, and the perpendicular EGI is drawn, the optical additosubtraction will be greatest around E and I. But the physical was greatest around D and T. Therefore the composite additosubtraction is greatest at the mean position between D and E and between T and I.

You have taught us how to compute the mean anomaly and the corrected anomaly from the assumption of anomaly of the eccentric: but practice more frequently requires us, given the mean anomaly, as from the time given, to find the others: teach this too.

Here there is no direct way; but he who wishes to compute this without tables must apply the rule of suppositions, namely by supposing that the anomaly of the eccentric—PK, in the accompanying diagram—is of this or of that magnitude, and upon that assumption, compute his mean anomaly PKA. For if PKA turns out to be as great as it was given to be, then the supposition as to PK the anomaly of the eccentric will be right. But if it does not turn out as great, then the supposition is to be corrected in view of the result, and the work done over again.

Could you by way of example teach a convenient method, lest through unfamiliarity one err too much in mistaken suppositions?

Then let the foregoing example be used again, and let the said mean anomaly [696] or area PKA be $50°9'10''$. It is manifest that, if the area of triangle KBA were known, the remaining area KBP would have the same number of degrees as its arc PK; and hence, if the value of KBA is subtracted from PKA, the remainder will be PK the anomaly of the eccentric. Therefore since PKA

is greater than *PKB*, the sine of arc *PX* will be smaller than the sine of 50° 9'10", and therefore smaller than 76.775. In our first supposition, let this sine be 70.000 for the sake of ease in multiplication.

Accordingly if this sine is multiplied by the value of triangle *DBA*—which was 11,910" in the preceding example—and the last five decimal places are cut off, the product *BKA* will be 8,337" or 2°18'57". Add these 2°18'57" to the arc of sine 70.000, which is 44°25'. The area *PKA* will be 46°44'. This is too small, for it is deficient in 3°25', since it ought to be 50°9', the quantity given.

Accordingly, in the second supposition, let a greater sine be supposed, by the addition of the deficiency of 3°25' to the 44°25' of the arc previously supposed, so that *PC* will be about 47°50'; its sine is approximately 74.000, which I choose for the sake of ease in calculation again. The multiplication of 74.000 by 11,910" makes *PKA* to be 7'56" greater, namely 2°26'53". Add this to the *PK* of the second supposition, namely to the 47°44'6" of *PKB*. Arc *PKA* will be 50°10'59"; and there will be an excess of 1'49" over the required 50°9'10".

And so we understand that this very small excess is to be subtracted from the third supposition of *PK*, and the remainder will be the required anomaly of the eccentric or *PK* will be equal to 47°42'17". It is possible to prove this. For the sine of this arc is 73.969, and the multiplication of that by 11,910" gives 2°26'50" for *KBA*. And the addition of that to 47°42'17" gives 50°9'7", which is imperceptibly different from the requisite 50°9'10".

5. On the Digression of the Planets Away from the Ecliptic

What is understood by the name "orbit"?

Properly speaking, it is that line which the planet describes around the sun

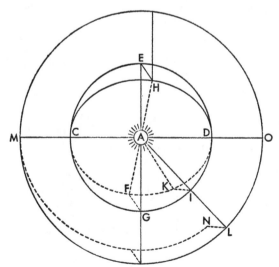

by means of the centre of its body. For example, in the diagram, if *ECGD* is a part of the plane of the ecliptic, *HCFD* will be the orbit.

But in a secondary sense, it is understood to be the [697] great circle wherein the plane of the orbit continued cuts the sphere of the fixed stars. For example, here it is section *MN* which has been made by plane *CAK* continued.

What do you call the "inclination of the planet" or "of any point in its orbit," and what is the "circle of inclination"?

Properly speaking, planets or points do not have an inclination, but lines or planes do. But because those planes are circumscribed around the orbits of the planets, and because the lines of movement of the planets are understood as having been described in those planes: in common usage these words have been transferred to the planets themselves, for the sake of brevity in speaking.

Therefore since that which below, in Book VI, will be called latitude, participates in that adventitious or optical irregularity—which is the second thing we will investigate—wherefore, in order that things which are distinct may also be distinguished by their names, let the real digression of the planet from the ecliptic be called, not latitude, but inclination. Now it is defined as being the arc of the great circle, in the sphere of the fixed stars, described around the sun, [698] perpendicular to the ecliptic and called the circle of inclination, which arc is intercepted between the ecliptic and the eccentric position of the planet. Or, it is the angle at the sun which this arc measures.

In the diagram, if A is the sun, FKDHC the orbit, and MLO the ecliptic, the inclination of point K will be angle KAI or NAL or its arc NL described around the sun A.

What do you call "nodes," and what, "limits"?

The nodes are the two points on the ecliptic wherein it is cut by the plane of the orbit continued. In Greek, σννδέσμοι, because in them different routes, the apparent route of the sun and that of the planet, are knotted together. The one is "the ascending node," where the planet leaves the southern hemisphere and turns northward; the other "the descending node," which transfers the planet to the south—the words "ascending" and "descending" having been accommodated to our hemisphere, as that in which the first founders of astronomy lived. For example, if the plane of the orbit and the plane of the ecliptic coincide in line CAD, which, on being prolonged into the sphere of the fixed stars, designates the ecliptic section, M and O will designate the nodes.

[699] But the points on the ecliptic which are 90° distant from the nodes are called the limits: the northern limit, wherefrom the planet is distant to the north, and the southern limit, wherefrom it is distant to the south. They are called limits, wherefrom it is distant to the south. They are called limits because the planet arriving at those points does not digress farther in any direction but from that place turns around and begins to return to the ecliptic. For example, in the diagram points E and G on the ecliptic are called the limits. But also points H and F on the real orbit and the corresponding points in the sphere of the fixed stars, are called by the same name, and more frequently too.

What do you call "the argument of the inclination"?

It is the arc of the planetary orbit in the sphere of the fixed stars intercepted between the ascending node and the eccentric position of the planet and measured eastwards. For example, if O is the ascending node, N the eccentric position of the planet, OMN will be the argument of inclination LN. Copernicus uses the northern limit instead of the ascending node.

The greatest inclination of the limit for any given planet is not the same throughout all the ages, is it?

According to the physical principles employed in Book IV, of itself it is unchangeable. But on account of the dislocation of the ecliptic—concerning which I shall speak in Book VII—it can change *per accidens*.

How is the inclination of the planet computed?

No differently from the way in which the declination of a point on the ecliptic was computed in Book III. If the sine of the greatest inclination is multipled by the sine of the argument of the inclination, and the last five decimal places are cut off from the product, the sine of the inclination will be given. See the method followed on folium 245 *et seqq.* If instead of the sines of the arcs you employ their logarithms, the multiplication will be converted to simple addition.

What is the eccentric position of the planet on the ecliptic?

That point on the ecliptic wherein it is cut by the circle of inclination drawn through the eccentric position [700] in the simple sense. For example, if the eccentric position—in the simple sense—of the planet at *K* is *N* and *NL* is the circle of inclination, and angles *NLM* and *NLO* are right, *L* will be the eccentric position of the planet on the ecliptic. It is not called the ecliptic position simply, because it also involves the second irregularity, the matter of Book VI; but the word "eccentric" is added in order for us to understand that it is a question of that position which is determined on the ecliptic by the eccentric alone, without the introduction of the great sphere, whereof I shall speak in Book VI.

What is the eccentric longitude of the planet held to be?

The arc of the ecliptic measured eastwards from the beginning of the Ram to the circle of inclination of the planet or its eccentric position on the ecliptic. It is called the eccentric longitude, not because it is measured on the eccentric but because the eccentric causes it.

What is meant by "the reduction to the ecliptic"?

The small arc which is equal to the difference between the argument of the inclination and the eccentric longitude—that is, the difference between the two arcs, one on the orbit and the other on the ecliptic, beginning at the common node and ending at the circle of inclination. For example, in this case it is the difference between *MN* and *ML*.

How is it computed?

Not otherwise than on page 255, Book III, the difference between the right ascension and the corresponding arc on the ecliptic is computed. For if the sine of the complement of the greatest inclination is multipled by the tangent of the argument of the inclination, and the last five decimal places are cut away from the product, the tangent of the argument of the reduction will be given.

Or else: the antilogarithm of the greatest inclination is added to the mesologarithm of the argument, and thus the sum will be the mesologarithm of the argument of the reduction.

A compendious and more useful method, even for the ascension, is as follows. The greatest reduction around 45° from the node, if multiplied by the sine of twice any arc, and the last five decimals cut off, constitute the required reduction for the arc taken simply.

[701] *How is this reduction to be employed, and in relation to what?*

When the planet is progressing from the nodes to the limits, the reduction is to be subtracted from the argument of the inclination; it is to be added, when the planet is progressing from the limits to the nodes; then the result, if added to the ascending node, constitutes the eccentric longitude of the position of the planet.

What do you call "the foreshortening"?

It is the particle of the distance of the planet from the centre of the sun which corresponds to the sagitta of the inclination of the planet in the same ratio in which the total interval corresponds to the total sine.

Let A be the sun, and P and Q the poles of the ecliptic. Let TAX represent the plane of the ecliptic and EAG the plane of the orbit. Let the planet now be at E or G and with A as centre and intervals AE, AG let the arcs EH and GF be drawn; and from E and G let the perpendiculars ER and GS be dropped upon TX. HR and SF will be the foreshortenings.

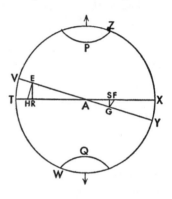

What is the foreshortened distance?

It is the straight line in the plane of the ecliptic between the centre of the sun and the perpendicular from the centre of the planetary body. In the diagram, if the planet is at E or G, then AR or AS is the foreshortened distance.

How is the foreshortened distance computed?

The assumed distance, expressed in the numbers of the measurement proper to each planet, is multipled by the sine of the complement of the inclination of the assumed distance, and the last five decimal places are removed from the product. Or, the logarithm [702] of the distance is added to the antilogarithms of the proper inclination, and the sum is the logarithm which is the index of the foreshortened distance.

Where is the distance foreshortened the most?

Around the limits, and more around the limit which is nearer the aphelion. For example, if V and Y are the limits, and thus Z and W are the poles of the orbit, and V is nearer the aphelion than Y, then HR will be longer than FS and will be the longest of all.

6. On the Movement of the Apsides and Nodes

How do you define the movement of the apsis among the primary planets?

It is the arc of the orbit in the sphere of the fixed stars intercepted between that point on the orbit which is at the same distance from the moving node as a fixed point on the ecliptic—*viz.*, the beginning of the Ram, or even the first star of the Ram—and the position of the highest apsis, and measured eastwards.

What kind of movement is this movement of the apsides?

It is set down as being uniform, (1) on account of its unbelievable slowness whereby the astronomers are impeded to such an extent that they cannot precisely investigate this movement in its single parts; (2) because we have an instance of regularity in one planet, where the period of the apsis of this physical movement, which we touched upon in Book IV, page 105, as resting upon mere conjectures, cannot be prejudicial to this regularity at all, although in virtue of them it seems that this movement can be made irregular. But more about this in Book VI, in connection with the single planets.

What is to be understood by the movement of the nodes in the primary planets, or what is the longitude of the node?

The movement of the node is the arc on the ecliptic measured westwards [703] from a fixed point on it—namely, either from the beginning of the Ram or from the position of the first star of the Ram—to the position of the ascending node. But if the measurement is made eastwards, then this arc too can be called the longitude of the node.

What kind of movement is this movement of the nodes?

Although it is reasonable that even the movement of this point is uniform of itself, nevertheless it seems that some irregularity is present in it *per accidens*, on account of the dislocation of the ecliptic, with which Book VII is concerned.

What figures do the movement of the nodes and limits describe?

The nodes proceed along the great circle of the ecliptic, while the limits of the orbit, in so far as their inclination is assumed to remain unchangeable, proceed in circles parallel to the ecliptic or to that circle with reference to which their inclination is unchangeable.

In order to aid the understanding, their movements can be represented by a

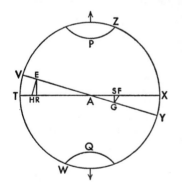

not absurd picture [of the movements] of the poles, provided we remember that, physically speaking, there is no need of the poles. For example, in the accompanying diagram, let the orbit be *VY*—by the continuation of the plane to the sphere of the fixed stars—and let its poles *Z* and *W* move in small circles around *P* and *Q* the poles of the ecliptic *TX*. Therefore in what direction away from *P* pole *Z* moves at any time, the limit *V* moves in this same direction away from point *T* on the ecliptic, and so does the limit *Y* away from point *X* on the ecliptic; and the circuit of *Z* in the small circle, which is parallel to *TX* in the same direction is followed by the limit *V* in northern parallel which is as much greater [than the small circle] as it is nearer to *TX*; and similarly *Y*, in the southern parallel. For the following six points will always be found in the great circle of inclination: *Z* and *W* the poles of the orbit, *P* and *Q* the poles of the ecliptic, and *V* and *Y* the limits of the orbit.

Accordingly, up to now we have been concerned with the definitions of the terms having to do with the planetary orbit and the eccentric circle circumscribed around it. Since they are common to all the five planets, they have been put ahead in Book v. However, we shall teach their employment, with respect to the single planets, in Book vi, which follows[1].

THE END OF BOOK FIVE, THE SECOND BOOK ON THE DOCTRINE OF THE SCHEMATA OR OF CELESTIAL PHYSICS

Harmonies of the World

Contents

167

BOOK FIVE

Concerning the very perfect harmony of the celestial movements, and the genesis of eccentricities and the semidiameters, and the periodic times from the same.

After the model of the most correct astronomical doctrine of today, and the hypothesis not only of Copernicus but also of Tycho Brahe, whereof either hypotheses are today publicly accepted as most true, and the Ptolemaic as outmoded.

I commence a sacred discourse, a most true hymn to God the Founder, and I judge it to be piety, not to sacrifice many hecatombs of bulls to Him and to burn incense of innumerable perfumes and cassia, but first to learn myself, and afterwards to teach others too, how great He is in wisdom, how great in power, and of what sort in goodness. For to wish to adorn in every way possible the things that should receive adornment and to envy no thing its goods—this I put down as the sign of the greatest goodness, and in this respect I praise Him as good that in the heights of His wisdom He finds everything whereby each thing may be adorned to the utmost and that He can do by his unconquerable power all that he has decreed.

GALEN, on the Use of Parts. BOOK III

PROEM

[268] As regards that which I prophesied two and twenty years ago (especially that the five regular solids are found between the celestial spheres), as regards that of which I was firmly persuaded in my own mind before I had seen Ptolemy's *Harmonies*, as regards that which I promised my friends in the title of this fifth book before I was sure of the thing itself, that which, sixteen years ago, in a published statement, I insisted must be investigated, for the sake of which I spent the best part of my life in astronomical speculations, visited Tycho Brahe, [269] and took up residence at Prague: finally, as God the Best and Greatest, Who had inspired my mind and aroused my great desire, prolonged my life and strength of mind and furnished the other means through the liberality of the two Emperors and the nobles of this province of Austria-on-the-Anisana: after I had discharged my astronomical duties as much as sufficed, finally, I say, I brought it to light and found it to be truer than I had even hoped, and I discovered among the celestial movements the full nature of harmony, in its due measure, together with all its parts unfolded in Book III—not in that mode wherein I had conceived it in my mind (this is not last in my joy) but in a very different mode which is also very excellent and very perfect. There took place in this intervening time, wherein the very laborious reconstruction of the movements held me in suspense, an extraordinary augmentation of my desire and incentive for the job, a reading of the *Harmonies* of Ptolemy, which had

been sent to me in manuscript by John George Herward, Chancellor of Bavaria, a very distinguished man and of a nature to advance philosophy and every type of learning. There, beyond my expectations and with the greatest wonder, I found approximately the whole third book given over to the same consideration of celestial harmony, fifteen hundred years ago. But indeed astronomy was far from being of age as yet; and Ptolemy, in an unfortunate attempt, could make others subject to despair, as being one who, like Scipio in Cicero, seemed to have recited a pleasant Pythagorean dream rather than to have aided philosophy. But both the crudeness of the ancient philosophy and this exact agreement in our meditations, down to the last hair, over an interval of fifteen centuries, greatly strengthened me in getting on with the job. For what need is there of many men? The very nature of things, in order to reveal herself to mankind, was at work in the different interpreters of different ages, and was the finger of God—to use the Hebrew expression; and here, in the minds of two men, who had wholly given themselves up to the contemplation of nature, there was the same conception as to the configuration of the world, although neither had been the other's guide in taking this route. But now since the first light eight months ago, since broad day three months ago, and since the sun of my wonderful speculation has shone fully a very few days ago: nothing holds me back. I am free to give myself up to the sacred madness, I am free to taunt mortals with the frank confession that I am stealing the golden vessels of the Egyptians, in order to build of them a temple for my God, far from the territory of Egypt. If you pardon me, I shall rejoice; if you are enraged, I shall bear up. The die is cast, and I am writing the book—whether to be read by my contemporaries or by posterity matters not. Let it await its reader for a hundred years, if God Himself has been ready for His contemplator for six thousand years.

The chapters of this book are as follows:

1. Concerning the five regular solid figures.
2. On the kinship between them and the harmonic ratios.
3. Summary of astronomical doctrine necessary for speculation into the celestial harmonies.
4. In what things pertaining to the planetary movements the simple consonances have been expressed and that all those consonances which are present in song are found in the heavens.
5. That the clefs of the musical scale, or pitches of the system, and the genera of consonances, the major and the minor, are expressed in certain movements.
6. That the single musical Tones or Modes are somehow expressed by the single planets.
7. That the counterpoints or universal harmonies of all the planets can exist and be different from one another.
8. That four kinds of voice are expressed in the planets: soprano, contralto, tenor, and bass.
[270] 9. Demonstration that in order to secure this harmonic arrangement, those very planetary eccentricities which any planet has as its own, and no others, had to be set up.
10. Epilogue concerning the sun, by way of very fertile conjectures.

Before taking up these questions, it is my wish to impress upon my readers the very exhortation of Timaeus, a pagan philosopher, who was going to speak on the same things: it should be learned by Christians with the greatest admiration, and shame too, if they do not imitate him: Ἀλλ᾽ ὦ Σώκρατες, τοῦτό γε δὴ πντες, ὅσοι καὶ κατὰ βραχὺ σωφροσύνης μετέχουσιν, ἐπὶ πασῇ ὁρμῇ καὶ σμικροῦ καὶ μεγάλου πράγματος θεὸν ἀει που καλοῦσιν. ἡμᾶς δὲ τοὺς περὶ τοῦ πάντος λόγους ποιεῖσθαι πῃ μέλλοντας . . . , εἰ μὴ παντάπασι παραλλάττομεν, ἀνάγκη θεοὺς τε καὶ θεὰς ἐπικαλουμένους εὔχεσθαι πάντα, κατὰ νοῦν ἐκείνοις μὲν μάλιστα, ἐπομένως δὲ ἡμῖν εἰπεῖν. *For truly, Socrates, since all who have the least particle of intelligence always invoke God whenever they enter upon any business, whether light or arduous; so too, unless we have clearly strayed away from all sound reason, we who intend to have a discussion concerning the universe must of necessity make our sacred wishes and pray to the Gods and Goddesses with one mind that we may say such things as will please and be acceptable to them in especial and, secondly, to you too.*

1. CONCERNING THE FIVE REGULAR SOLID FIGURES

[271] It has been said in the second book how the regular plane figures are fitted together to form solids; there we spoke of the five regular solids, among others, on account of the plane figures. Nevertheless their number, five, was there demonstrated; and it was added why they were designated by the Platonists as the figures of the world, and to what element any solid was compared on account of what property. But now, in the anteroom of this book, I must speak again concerning these figures, on their own account, not on account of the planes, as much as suffices for the celestial harmonies; the reader will find the rest in the *Epitome of Astronomy*, Volume II, Book IV.

Accordingly, from the *Mysterium Cosmographicum*, let me here briefly inculcate the order of the five solids in the world, whereof three are primary and two secondary. For the *cube* (1) is the outmost and the most spacious, because firstborn and having the nature [*rationem*] of a *whole*, in the very form of its generation. There follows the *tetrahedron* (2), as if made a *part*, by cutting up the cube; nevertheless it is primary too, with a solid trilinear angle, like the cube. Within the tetrahedron is the *dodecahedron* (3), the last of primary figures, namely, like a solid composed of parts of a cube and similar parts of a tetrahedron, *i.e.*, of irregular tetrahedrons, wherewith the cube inside is roofed over. Next in order is the *icosahedron* (4) on account of its similarity, the last of the secondary figures and having a plurilinear solid angle. The *octahedron* (5) is inmost, which is similar to the cube and the first of the secondary figures and to which as inscriptile the first place is due, just as the first outside place is due to the cube as circumscriptile.

[272] However, there are as it were two noteworthy weddings of these figures, made from different classes: the males, the cube and the dodecahedron, among the primary; the females, the octahedron and the icosahedron, among the secondary, to which is added one as it were bachelor or hermaphrodite, the tetrahedron, because it is inscribed in itself, just as those female solids are inscribed in the males and are as it were subject to them, and have the signs of the feminine sex, opposite the masculine, namely, angles opposite planes. Moreover, just as the tetrahedron is the element, bowels, and as it were rib of the male

cube, so the feminine octahedron is the element and part of the tetrahedron in another way; and thus the tetrahedron mediates in this marriage.

The main difference in these wedlocks or family relationships consists in the following: the ratio of the cube is *rational*. For the tetrahedron is one third of the body of the cube, and the octahedron half of the tetrahedron, one sixth of the cube; while the ratio of the dodecahedron's wedding is *irrational* [*ineffabilis*] but *divine*.

The union of these two words commands the reader to be careful as to their significance. For the word *ineffabilis* here does not of itself denote any nobility, as elsewhere in theology and divine things, but denotes an inferior condition. For in geometry, as was said in the first book, there are many irrationals, which do not on that account participate in a divine proportion too. But you must look in the first book for what the divine ratio, or rather the divine section, is. For in other proportions there are four terms present; and three, in a continued proportion; but the divine requires a single relation of terms outside of that of the proportion itself, namely in such fashion that the two lesser terms, as parts make up the greater term, as a whole. Therefore, as much as is taken away from this wedding of the dodecahedron on account of its employing an irrational proportion, is added to it conversely, because its irrationality approaches the divine. This wedding also comprehends the solid star too, the generation whereof arises from the continuation of five planes of the dodecahedron till they all meet in a single point. See its generation in Book II.

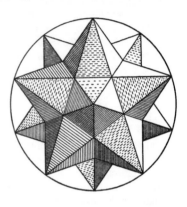

Lastly, we must note the ratio of the spheres circumscribed around them to those inscribed in them: in the case of the tetrahedron it is rational, 100,000 : 33,333 or 3 : 1; in the wedding of the cube it is irrational, but the radius of the inscribed sphere is rational in square, and is itself the square root of one third the square on the radius [of the circumscribed sphere], namely 100,000 : 57,735; in the wedding of the dodecahedron, clearly irrational, 100,000 : 79,465; in the case of the star, 100,000 : 52,573, half the side of the icosahedron or half the distance between two rays.

2. ON THE KINSHIP BETWEEN THE HARMONIC RATIOS AND THE FIVE REGULAR FIGURES

[273] This kinship [*cognatio*] is various and manifold; but there are four degrees of kinship. For either the sign of kinship is taken from the outward form alone which the figures have, or else ratios which are the same as the harmonic arise in the construction of the side, or result from the figures already constructed, taken simply or together; or, lastly, they are either equal to or approximate the ratios of the spheres of the figure.

In the first degree, the ratios, where the character or greater term is 3, have kinship with the triangular plane of the tetrahedron, octahedron, and icosahedron; but where the greater term is 4, with the square plane of the cube; where 5, with the pentagonal plane of the dodecahedron. This similitude on the part of the plane can also be extended to the smaller term of the ratio, so that wherever the number 3 is found as one term of the continued doubles, that ratio is held to be akin to the three figures first named: for example, 1 : 3 and 2 : 3 and 4 : 3 and 8 : 3, et cetera; but where the number is 5, that ratio is abolutely assigned to the wedding of the dodecahedron: for example, 2 : 5 and 4 : 5 and 8 : 5, and thus 3 : 5 and 3 : 10 and 6 : 5 and 12 : 5 and 24 : 5. The kinship will be less probable if the sum of the terms expresses this similitude, as in 2 : 3 the sum of the terms is equal to 5, as if to say that 2 : 3 is akin to the dodecahedron. The kinship on account of the outward form of the solid angle is similar: the solid angle is trilinear among the primary figures, quadrilinear in the octahedron, and quinquelinear in the icosahedron. And so if one term of the ratio participates in the number 3, the ratio will be connected with the primary bodies; but if in the number 4, with the octahedron; and finally, if in the number 5, with the icosahedron. But in the feminine solids this kinship is more apparent, because the characteristic figure latent within follows upon the form of the angle: the tetragon in the octahedron, the pentagon in the icosahedron; and so 3 : 5 would go to the sectioned icosahedron for both reasons.

The second degree of kinship, which is genetic, is to be conceived as follows: First, some harmonic ratios of numbers are akin to one wedding or family, namely, perfect ratios to the single family of the cube; conversely, there is the ratio which is never fully expressed in numbers and cannot be demonstrated by numbers in any other way, except by a long series of numbers gradually approaching it: this ratio is called *divine*, when it is perfect, and it rules in various ways throughout the dodecahedral wedding. Accordingly, the following consonances begin to shadow forth that ratio: 1 : 2 and 2 : 3 and 2 : 3 and 5 : 8. For it exists most imperfectly in 1 : 2, more perfectly in 5 : 8, and still more perfectly if we add 5 and 8 to make 13 and take 8 as the numerator, if this ratio has not stopped being harmonic.

Further, in constructing the side of the figure, the diameter of the globe must be cut; and the octahedron demands its bisection, the cube and the tetrahedron its trisection, the dodecahedral wedding its quinquesection. Accordingly, the ratios between the figures are distributed according to the numbers which express those ratios. But the square on the diameter is cut too, or the square on the side of the figure is formed from a fixed part of the diameter. And then the squares on the sides are compared with the square on the diameter, and they constitute the following ratios: in the cube 1 : 3, in the tetrahedron 2 : 3, in the octahedron 1 : 2. Wherefore, if the two ratios are put together, the cubic and the tetrahedral will give 1 : 2; the cubic and the octahedral, 2 : 3; the octahedral and the tetrahedral, 3 : 4. The sides in the dodecahedral wedding are irrational.

Thirdly, the harmonic ratios follow in various ways upon the already constructed figures. For either the number of the sides of the plane is compared with the number of lines in the total figure; [274] and the following ratios arise: in the cube, 4 : 12 or 1 : 3; in the tetrahedron 3 : 6 or 1 : 2; in the octahedron 3 : 12 or 1 : 4; in the dodecahedron 5 : 30 or 1 : 6; in the icosahedron 3 : 30 or 1 : 10. Or else the number of sides of the plane is compared with the number of planes; then the cube gives 4 : 6 or 2 : 3, the tetrahedron 3 : 4, the octahedron 3 : 8, the dodecahedron 5 : 12, the icosahedron 3 : 20. Or else the number of sides or angles of the plane is compared with the number of solid angles, and the cube gives 4 : 8 or 1 : 2, the tetrahedron 3 : 4, the octahedron 3 : 6 or 1 : 2, the dodecahedron with its consort 5 : 20 or 3 : 12 (*i.e.*, 1 : 4). Or else the number of planes is compared with the number of solid angles, and the cubic wedding gives 6 : 8 or 3 : 4, the tetrahedron the ratio of equality, the dodecahedral wedding 12 : 20 or 3 : 5. Or else the number of all the sides is compared with the number of the solid angles, and the cube gives 8 : 12 or 2 : 3, the tetrahedron 4 : 6 or 2 : 3, and the octahedron 6 : 12 or 1 : 2, the dodecahedron 20 : 30 or 2 : 3, the icosahedron 12 : 30 or 2 : 5.

Moreover, the bodies too are compared with one another, if the tetrahedron is stowed away in the cube, the octahedron in the tetrahedron and cube, by geometrical inscription. The tetrahedron is one third of the cube, the octahedron half of the tetrahedron, one sixth of the cube, just as the octahedron, which is inscribed in the globe, is one sixth of the cube which circumscribes the globe. The ratios of the remaining bodies are irrational.

The fourth species or degree of kinship is more proper to this work: the ratio of the spheres inscribed in the figures to the spheres circumscribing them is sought, and what harmonic ratios approximate them is calculated. For only in the tetrahedron is the diameter of the inscribed sphere rational, namely, one third of the circumscribed sphere. But in the cubic wedding the ratio, which is single there, is as lines which are rational only in square. For the diameter of the inscribed sphere is to the diameter of the circumscribed sphere as the square root of the ratio 1 : 3. And if you compare the ratios with one another, the ratio of the tetrahedral spheres is the square of the ratio of the cubic spheres. In the dodecahedral wedding there is again a single ratio, but an irrational one, slightly greater than 4 : 5. Therefore the ratio of the spheres of the cube and octahedron is approximated by the following consonances: 1 : 2, as proximately greater, and 3 : 5, as proximately smaller. But the ratio of the dodecahedral spheres is approximated by the consonances 4 : 5 and 5 : 6, as proximately smaller, and 3 : 4 and 5 : 8, as proximately greater.

But if for certain reasons 1 : 2 and 1 : 3 are arrogated to the cube, the ratio of the spheres of the cube will be to the ratio of the spheres of the tetrahedron as the consonances 1 : 2 and 1 : 3, which have been ascribed to the cube, are to 1 : 4 and 1 : 9, which are to be assigned to the tetrahedron, if this proportion is to be used. For these ratios, too, are as the squares of those consonances. And because 1 : 9 is not harmonic, 1 : 8 the proximate ratio takes its place in the tetrahedron. But by this proportion approximately 4 : 5 and 3 : 4 will go with the dodecahedral wedding. For as the ratio of the spheres of the cube is approximately the cube of the ratio of the dodecahedral, so too the cubic consonances 1 : 2 and 2 : 3 are approximately the cubes of the consonances 4 : 5 and 3 : 4. For 4 : 5 cubed is 64 : 125, and 1 : 2 is 64 : 128. So 3 : 4 cubed is 27 : 64, and 1 : 3 is 27 : 81.

3. A Summary of Astronomical Doctrine Necessary for Speculation into the Celestial Harmonies

First of all, my readers should know that the ancient astronomical hypotheses of Ptolemy, in the fashion in which they have been unfolded in the *Theoricae* of Peurbach and by the other writers of epitomes, are to be completely removed from this discussion and cast out of [275] the mind. For they do not convey the true lay out of the bodies of the world and the polity of the movements.

Although I cannot do otherwise than to put solely Copernicus' opinion concerning the world in the place of those hypotheses and, if that were possible, to persuade everyone of it; but because the thing is still new among the mass of the intelligentsia [apud vulgus studiosorum], and the doctrine that the Earth is one of the planets and moves among the stars around a motionless sun sounds very absurd to the ears of most of them: therefore those who are shocked by the unfamiliarity of this opinion should know that these harmonical speculations are possible even with the hypotheses of Tycho Brahe—because that author holds, in common with Copernicus, everything else which pertains to the lay out of the bodies and the tempering of the movements, and transfers solely the Copernican annual movement of the Earth to the whole system of planetary spheres and to the sun, which occupies the centre of that system, in the opinion of both authors. For after this transference of movement it is nevertheless true that in Brahe the Earth occupies at any time the same place that Copernicus gives it, if not in the very vast and measureless region of the fixed stars, at least in the system of the planetary world. And accordingly, just as he who draws a circle on paper makes the writing-foot of the compass revolve, while he who fastens the paper or tablet to a turning lathe draws the same circle on the revolving tablet with the foot of the compass or stylus motionless; so too, in the case of Copernicus the Earth, by the real movement of its body, measures out a circle revolving midway between the circle of Mars on the outside and that of Venus on the inside; but in the case of Tycho Brahe the whole planetary system (wherein among the rest the circles of Mars and Venus are found) revolves like a tablet on a lathe and applies to the motionless Earth, or to the stylus on the lathe, the midspace between the circles of Mars and Venus; and it comes about from this movement of the system that the Earth within it, although remaining motionless, marks out the same circle around the sun and midway between Mars and Venus, which in Copernicus it marks out by the real movement of its body while the system is at rest. Therefore, since harmonic speculation considers the eccentric movements of the planets, as if seen from the sun, you may easily understand that if any observer were stationed on a sun as much in motion as you please, nevertheless for him the Earth, although at rest (as a concession to Brahe), would seem to describe the annual circle midway between the planets and in an intermediate length of time. Wherefore, if there is any man of such feeble wit that he cannot grasp the movement of the earth among the stars, nevertheless he can take pleasure in the most excellent spectacle of this most divine construction, if he applies to their image in the sun whatever he hears concerning the daily movements of the Earth in its eccentric—such an image as Tycho Brahe exhibits, with the Earth at rest.

And nevertheless the followers of the true Samian philosophy have no just cause to be jealous of sharing this delightful speculation with such persons, because their joy will be in many ways more perfect, as due to the consummate perfection of speculation, if they have accepted the immobility of the sun and the movement of the earth.

Firstly [I], therefore, let my readers grasp that today it is absolutely certain among all astronomers that all the planets revolve around the sun, with the exception of the moon, which alone has the Earth as its centre: the magnitude of the moon's sphere or orbit is not great enough for it to be delineated in this diagram in a just ratio to the rest. Therefore, to the other five planets, a sixth, the Earth, is added, which traces a sixth circle around the sun, whether by its own proper movement with the sun at rest, or motionless itself and with the whole planetary system revolving.

Secondly [II]: It is also certain that all the planets are eccentric, *i.e.*, they change their distances from the sun, in such fashion that in one part of their circle they become farthest away from the sun, [276] and in the opposite part they come nearest to the sun. In the accompanying diagram three circles apiece have been drawn for the single planets: none of them indicate the eccentric route of the planet itself; but the mean circle, such as *BE* in the case of Mars, is equal to the eccentric orbit, with respect to its longer diameter. But the orbit itself, such as *AD*, touches *AF*, the upper of the three, in one place *A*, and the lower circle *CD*, in the opposite place *D*. The circle *GH* made with dots and described through the centre of the sun indicates the route of the sun according to Tycho Brahe. And if the sun moves on this route, then absolutely all the points in this whole planetary system here depicted advance upon an equal route, each upon his own. And with one point of it (namely, the centre of the sun) stationed at one point of its circle, as here at the lowest, absolutely each and every point of the system will be stationed at the lowest part of

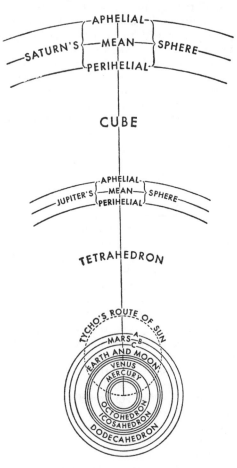

its circle. However, on account of the smallness of the space the three circles of Venus unite in one, contrary to my intention.

Thirdly [III]: Let the reader recall from my *Mysterium Cosmographicum*, which I published twenty-two years ago, that the number of the planets or circular routes around the sun was taken by the very wise Founder from the five regular solids, concerning which Euclid, so many ages ago, wrote his book which is called the *Elements* in that it is built up out of a series of propositions. But it has been made clear in the second book of this work that there cannot be more regular bodies, *i.e.*, that regular plane figures cannot fit together in a solid more than five times.

Fourthly [IV]: As regards the ratio of the planetary orbits, the ratio between two neighbouring planetary orbits is always of such a magnitude that it is easily apparent that each and every one of them approaches the single ratio of the spheres of one of the five regular solids, namely, that of the sphere circumscribing to the sphere inscribed in the figure. Nevertheless it is not wholly equal, as I once dared to promise concerning the final perfection of astronomy. For, after completing the demonstration of the intervals from Brahe's observations, I discovered the following: if the angles of the cube [277] are applied to the inmost circle of Saturn, the centres of the planes are approximately tangent to the middle circle of Jupiter; and if the angles of the tetrahedron are placed against the inmost circle of Jupiter, the centres of the planes of the tetrahedron are approximately tangent to the outmost circle of Mars; thus if the angles of the octahedron are placed against any circle of Venus (for the total interval between the three has been very much reduced), the centres of the planes of the octahedron penetrate and descend deeply within the outmost circle of Mercury, but nonetheless do not reach as far as the middle circle of Mercury; and finally, closest of all to the ratios of the dodecahedral and icosahedral spheres—which ratios are equal to one another—are the ratios or intervals between the circles of Mars and the Earth, and the Earth and Venus; and those intervals are similarly equal, if we compute from the inmost circle of Mars to the middle circle of the Earth, but from the middle circle of the Earth to the middle circle of Venus. For the middle distance of the Earth is a mean proportional between the least distance of Mars and the middle distance of Venus. However, these two ratios between the planetary circles are still greater than the ratios of those two pairs of spheres in the figures, in such fashion that the centres of the dodecahedral planes are not tangent to the outmost circle of the Earth, and the centres of the icosahedral planes are not tangent to the outmost circle of Venus; nor, however, can this gap be filled by the semidiameter of the lunar sphere, by adding it, on the upper side, to the greatest distance of the Earth and subtracting it, on the lower, from the least distance of the same. But I find a certain other ratio of figures—namely, if I take the augmented dodecahedron, to which I have given the name of echinus, (as being fashioned from twelve quinquangular stars and thereby very close to the five regular solids), if I take it, I say, and place its twelve points in the inmost circle of Mars, then the sides of the pentagons, which are the bases of the single rays or points, touch the middle circle of Venus. In short: the cube and the octahedron, which are consorts, do not penetrate their planetary spheres at all; the dodecahedron and the icosahedron, which are consorts, do not wholly reach to theirs, the tetrahedron exactly touches both: in the first case there is falling short; in the second, excess; and in the third, equality, with respect to the planetary intervals.

Wherefore it is clear that the very ratios of the planetary intervals from the sun have not been taken from the regular solids alone. For the Creator, who is the very source of geometry and, as Plato wrote, "practices eternal geometry," does not stray from his own archetype. And indeed that very thing could be inferred from the fact that all the planets change their intervals throughout fixed periods of time, in such fashion that each has two marked intervals from the sun, a greatest and a least; and a fourfold comparison of the intervals from the sun is possible between two planets: the comparison can be made between either the greatest, or the least, or the contrary intervals most remote from one another, or the contrary intervals nearest together. In this way the comparisons made two by two between neighbouring planets are twenty in number, although on the contrary there are only five regular solids. But it is consonant that if the Creator had any concern for the ratio of the spheres in general, He would also have had concern for the ratio which exists between the varying intervals of the single planets specifically and that the concern is the same in both cases and the one is bound up with the other. If we ponder that, we will comprehend that for setting up the diameters and eccentricities conjointly, there is need of more principles, outside of the five regular solids.

Fifthly [v]: To arrive at the movements between which the consonances have been set up, once more I impress upon the reader that in the *Commentaries on Mars* I have demonstrated from the sure observations of Brahe that daily arcs, which are equal in one and the same eccentric circle, are not traversed with equal speed; but that these differing *delays in equal parts of the eccentric observe the ratio of their distances from the sun*, the source of movement; and conversely, that if equal times are assumed, namely, one natural day in both cases, the corresponding *true diurnal arcs* [278] *of one eccentric orbit have to one another the ratio which is the inverse of the ratio of the two distances from the sun.* Moreover, I demonstrated at the same time that *the planetary orbit is elliptical and the sun, the source of movement, is at one of the foci of this ellipse; and so, when the planet has completed a quarter of its total circuit from its aphelion, then it is exactly at its mean distance from the sun, midway between its greatest distance at the aphelion and its least at the perihelion.* But from these two axioms it results *that the diurnal mean movement of the planet in its eccentric is the same as the true diurnal arc of its eccentric at those moments wherein the planet is at the end of the quadrant of the eccentric measured from the aphelion, although that true quadrant appears still smaller than the just quadrant.* Furthermore, it follows *that the sum of any two true diurnal eccentric arcs, one of which is at the same distnace from the aphelion that the other is from the perihelion, is equal to the sum of the two mean diurnal arcs.* And as a consequence, *since the ratio of circles is the same as that of the diameters, the ratio of one mean diurnal arc to the sum of all the mean and equal arcs in the total circuit is the same as the ratio of the mean diurnal arc to the sum of all the true eccentric arcs, which are the same in number but unequal to one another.* And those things should first be known concerning the true diurnal arcs of the eccentric and the true movements, so that by means of them we may understand the movements which would be apparent if we were to suppose an eye at the sun.

Sixthly [vi]: But as regards the arcs which are apparent, as it were, from the sun, it is known even from the ancient astronomy that, among true movements which are equal to one another, that movement which is farther distant from the centre of the world (as being at the aphelion) will appear smaller to a beholder at that centre, but the movement which is nearer (as being at the perihelion) will similarly appear greater. Therefore, since moreover the true diurnal arcs at the near distance are still greater, on account of the faster movement, and still smaller at the distant aphelion, on account of the slowness of the movement, I demonstrated in the *Commentaries on Mars* that *the ratio of the apparent diurnal arcs of one eccentric circle is fairly exactly the inverse ratio of the squares of their distances from the sun.* For example, if the planet one day when it is at a distance from the sun of 10 parts, in any measure whatsoever, but on the opposite day, when it is at the perihelion, of 9 similar parts: it is certain that from the sun its apparent progress at the aphelion will be to its apparent progress at the perihelion, as 81 : 100.

But that is true with these provisos: First, that the eccentric arcs should not be great, lest they partake of distinct distances which are very different—*i.e.*, lest the distances of their termini from the apsides cause a perceptible variation; second, that the eccentricity should not be very great, for the greater its eccentricity (*viz.*, the greater the arc becomes) the more the angle of its apparent movement increases beyond the measure of its approach to the sun, by Theorem 8 of Euclid's *Optics*; none the less in small arcs even a great distance is of no moment, as I have remarked in my *Optics*, Chapter 11. But there is another reason why I make that admonition. For the eccentric arcs around the mean anomalies are viewed obliquely from the centre of the sun. This obliquity subtracts from the magnitude of the apparent movement, since conversely the arcs around the apsides are presented directly to an eye stationed as it were at the sun. Therefore, when the eccentricity is very great, then the eccentricity takes away perceptibly from the ratio of the movements; if without any diminution we apply the mean diurnal movement to the mean distance, as if at the mean distance, it would appear to have the same magnitude which it does have—as will be apparent below in the case of Mercury. All these things are treated at greater length in Book v of the *Epitome of Copernican Astronomy*; but they have been mentioned here too because they have to do with the very terms of the celestial consonances, considered in themselves singly and separately.

Seventhly [VII]: If by chance anyone runs into those diurnal movements which are apparent [279] to those gazing not as it were from the sun but from the Earth, with which movements Book VI of the *Epitome of Copernican Astronomy* deals, he should know that their rationale is plainly not considered in this business. Nor should it be, since the Earth is not the source of the planetary movements, nor can it be, since with respect to deception of sight they degenerate not only into mere quiet or apparent stations but even into retrogradation, in which way a whole infinity of ratios is assigned to all the planets, simultaneously and equally. Therefore, in order that we may hold for certain what sort of ratios of their own are constituted by the single real eccentric orbits (although these too are still apparent, as it were to one looking from the sun, the source of movement), first we must remove from those movements of their own this image of the adventitious annual movement common to all five, whether it arises from the movement of the Earth itself, according to Copernicus, or from the annual movement of the total system, according to Tycho Brahe, and the winnowed movements proper to each planet are to be presented to sight.

Eighthly [VIII]: So far we have dealt with the different delays or arcs of one and the same planet. Now we must also deal with the comparison of the movements of two planets. Here take note of the definitions of the terms which will be necessary for us. We give the name of *nearest apsides* of two planets to the perihelion of the upper and the aphelion of the lower, notwithstanding that they tend not towards the same region of the world but towards distinct and perhaps contrary regions. By *extreme movements* understand the slowest and the fastest of the whole planetary circuit; by *converging or converse extreme movements*, those which are at the nearest apsides of two planets—namely, at the perihelion of the upper planet and the aphelion of the lower; by *diverging or diverse*, those at the opposite apsides—namely, the aphelion of the upper and the perihelion of the lower. Therefore again, a certain part of my *Mysterium Cosmographicum*, which was suspended twenty-two years ago, because it was not yet clear, is to be completed and herein inserted. For after finding the true intervals of the spheres by the observations of Tycho Brahe and continuous labour and much time, at last, at last the right ratio of the periodic times to the spheres

> *though it was late, looked to the unskilled man,*
> *yet looked to him, and, after much time, came,*

and, if you want the exact time, was conceived mentally on the 8th of March in this year One Thousand Six Hundred and Eighteen but unfelicitously submitted to calculation and rejected as false, finally, summoned back on the 15th of May, with a fresh assault undertaken, outfought the darkness of my mind by the great proof afforded by my labor of seventeen years on Brahe's observations and meditation upon it uniting in one concord, in such fashion that I first believed I was dreaming and was presupposing the object of my search among the principles. But it is absolutely certain and exact that *the ratio which exists between the periodic times of any two planets is precisely the ratio of the $\frac{3}{2}$th power of the mean distances*, i.e., *of the spheres themselves*; provided, however, that the arithmetic mean between both diameters of the elliptic orbit be slightly less than the longer diameter. And so if any one take the period, say, of the Earth, which is one year, and the period of Saturn, which is thirty years, and extract the cube roots of this ratio and then square the ensuing ratio by squaring the cube roots, he will have as his numerical products the most just ratio of the distances of the Earth and Saturn from the sun.[1] For the cube root of 1 is 1, and the square of it is 1; and the cube root of 30 is greater than 3, and therefore the square of it is greater than 9. And Saturn, at its mean distance from the sun, is slightly higher [280] than nine times the mean distance of the Earth from the sun. Further on, in Chapter 9, the use of this theorem will be necessary for the demonstration of the eccentricities.

Ninthly [IX]: If now you wish to measure with the same yardstick, so to speak, the true daily journeys of each planet through the ether, two ratios are to be compounded—the ratio of the true (not the apparent) diurnal arcs of the eccentric, and the ratio of the mean intervals of each planet from the sun (because that is the same as the ratio of the amplitude of the spheres), *i.e.*, *the true diurnal arc of each planet is to be multiplied by the semidiameter of its sphere*: the products will be numbers fitted for investigating whether or not those journeys are in harmonic ratios.

Tenthly [X]: In order that you may truly know how great any one of these diurnal journeys appears to be to an eye stationed as it were at the sun, although this same thing can be got immediately from the astronomy, nevertheless it will also be manifest if you multiply the ratio of the journeys by the inverse ratio not of the mean, but of the true intervals which exist at any position on

---

[1]For in the *Commentaries on Mars*, chapter 48, page 232, I have proved that this Arithmetic mean is either the diameter of the circle which is equal in length to the elliptic orbit, or else is very slightly less.

the eccentrics: *multiply the journey of the upper by the interval of the lower planet from the sun, and conversely multiply the journey of the lower by the interval of the upper from the sun.*

Eleventhly [XI]: And in the same way, if the apparent movements are given, at the aphelion of the one and at the perihelion of the other, or conversely or alternately, the ratios of the distances of the aphelion of the one to the perihelion of the other may be elicited. But where the mean movements must be known first, *viz.*, the inverse ratio of the periodic times, wherefrom the ratio of the spheres is elicited by Article VIII above: then *if the mean proportional between the apparent movement of either one of its mean movement be taken, this mean proportional is to the semidiameter of its sphere* (which is already known) *as the mean movement is to the distance or interval sought.* Let the periodic times of two planets be 27 and 8. Therefore the ratio of the mean diurnal movement of the one to the other is 8 : 27. Therefore the semidiameters of their spheres will be as 9 to 4. For the cube root of 27 is 3, that of 8 is 2, and the squares of these roots, 3 and 2, are 9 and 4. Now let the apparent aphelial movement of the one be 2 and the perihelial movement of the other $33\frac{1}{3}$. The mean proportionals between the mean movements 8 and 27 and these apparent ones will be 4 and 30. Therefore if the mean proportional 4 gives the mean distance of 9 to the planet, then the mean movement of 8 gives an aphelial distance 18, which corresponds to the apparent movement 2; and if the other mean proportional 30 gives the other planet a mean distance of 4, then its mean movement of 27 will give it a perihelial interval of $3\frac{3}{5}$. I say, therefore, that the aphelial distance of the former is to the perihelial distance of the latter as 18 to $3\frac{3}{5}$. Hence it is clear that if the consonances between the extreme movements of two planets are found and the periodic times are established for both, the extreme and the mean distances are necessarily given, wherefore also the eccentricities.

Twelfthly [XII]: It is also possible, from the different extreme movements of one and the same planet, to find the *mean movement.* The mean movement is not exactly the arithmetic mean between the extreme movements, nor exactly the geometric mean, but it is as much less than the geometric mean as the geometric mean is less than the [arithmetic] mean between both means. Let the two extreme movements be 8 and 10: the mean movement will be less than 9, and also less than the square root of 80 by half the difference between 9 and the square root of 80. In this way, if the aphelial movement is 20 and the perihelial 24, the mean movement will be less than 22, even less than the square root of 480 by half the difference between that root and 22. There is use for this theorem in what follows.

[281] Thirteenthly [XIII]: From the foregoing the following proposition is demonstrated, which is going to be very necessary for us: Just as the ratio of the mean movements of two planets is the inverse ratio of the $\frac{3}{2}$th powers of the spheres, so the ratio of two apparent converging extreme movements always falls short of the ratio of the $\frac{3}{2}$th powers of the intervals corresponding to those extreme movements; and in what ratio the product of the two ratios of the corresponding intervals to the two mean intervals or to the semidiameters of the two spheres falls short of the ratio of the square roots of the spheres, in that ratio does the ratio of the two extreme converging movements exceed the ratio of the corresponding intervals; but if that compound ratio were to exceed the

ratio of the square roots of the spheres, then the ratio of the converging movements would be less than the ratio of their intervals.[1]

Let the ratio of the spheres be $DH : AE$; let the ratio of the mean movements be $HI : EM$, the $\frac{3}{2}$th power of the inverse of the former.
Let the least interval of the sphere of the first be CG; and the greatest interval of the sphere of the second be BF; and first let $DH : CG$ comp. $BF : AE$ be smaller than the $\frac{1}{2}$th power of $DH : AE$. And let GH be the apparent perihelial movement of the upper planet, and FL the aphelial of the lower, so that they are converging extreme movements.

I say that
$$GK : FL = BF : CG$$
$$BF^{\frac{3}{4}} : CG^{\frac{3}{4}}.$$

For
$$HI : GK = CG^2 : DH^2;$$

and
$$FL : EM = AE^2 : BF^2.$$

Hence
$$HI : GK \text{ comp. } FL : EM = CG^2 : DH^2 \text{ comp. } AE^2 : BF^2.$$

But
$$CG : DH \text{ comp. } AE : BF < AE^{\frac{1}{4}} : DH^{\frac{1}{4}}$$
by a fixed ratio of defect, as was assumed. Therefore too
$$HI : GK \text{ comp. } FL : EM \quad AE^{\frac{3}{4}} : DH^{\frac{3}{4}}$$
$$AE : DH$$
by a ratio of defect which is the square of the former. But by number VIII
$$HI : EM = AE^{\frac{3}{4}} : DH^{\frac{3}{4}}.$$
Therefore let the ratio which is smaller by the total square of the ratio of defect be divided into the ratio of the $\frac{3}{2}$th powers; that is,
$$HI : EM \text{ comp. } GK : HI \text{ comp. } EM : FL \quad AE^{\frac{1}{4}} : DH^{\frac{1}{4}}$$
by the excess squared. But
$$HI : EM \text{ comp. } GK : HI \text{ comp. } EM : FL = GK : FL.$$
Therefore
$$GK : FL \quad AE^{\frac{1}{4}} : DH^{\frac{1}{4}}$$
by the excess squared. But
$$AE : DH = AE : BF \text{ comp. } BF : CG \text{ comp. } CG : DH.$$
And
$$CG : DH \text{ comp. } AE : BF \quad AE^{\frac{1}{4}} : DH^{\frac{1}{4}}$$
by the simple defect. Therefore
$$BF : CG \quad AE^{\frac{1}{4}} : DH^{\frac{1}{4}}$$
by the simple excess. But
$$GK : FL \quad AE^{\frac{1}{4}} : DH^{\frac{1}{4}}$$
but by the excess squared. But the excess squared is greater than the simple excess. Therefore the ratio of the movements GK to FL is greater than the ratio of the corresponding intervals BF to CG.

---

[1]Kepler always measures the magnitude of a ratio from the greater term to the smaller, rather than from the antecedent to the consequent, as we do today. For example, as Kepler speaks, 2 : 3 is the same as 3 : 2, and 3 : 4 is greater than 7 : 8.—C. G. Wallis.

In fully the same way, it is demonstrated even contrariwise that if the planets approach one another in G and F beyond the mean distances in H and E, in such fashion that the ratio of the mean distances $DH : AE$ becomes less than $DH^{\frac{1}{2}} : AE^{\frac{1}{2}}$, then the ratio of the movements $GK : FL$ becomes less than the ratio of the corresponding intervals $BF : CG$. For you need to do nothing more than to change the words *greater* to *less*, $>$to$<$, *excess* to *defect*, and conversely.

In suitable numbers, because the square root of $\frac{4}{9}$ is $\frac{2}{3}$; and $\frac{5}{8}$ is even greater than $\frac{2}{3}$ by the ratio of excess $15\frac{}{16}$; and the square of the ratio 8 : 9 [282] is the ratio 1600 : 2025, *i.e.*, 64 : 81; and the square of the ratio 4 : 5 is the ratio 3456 : 5400, *i.e.*, 16 : 25; and finally the $\frac{3}{2}$th power of the ratio 4 : 9 is the ratio 1600 : 5400, *i.e.*, 8 : 27: therefore too the ratio 2025 : 3456, *i.e.*, 75 : 128, is even greater than 5 : 8, *i.e.*, 75 : 120, by the same ratio of excess (*i.e.*, 120 : 128), 15 : 16; whence 2025 : 3456, the ratio of the converging movements, exceeds 5 : 8, the inverse ratio of the corresponding intervals, by as much as 5 : 8 exceeds 2 : 3, the square root of the ratio of the spheres. Or, what amounts to the same thing, the ratio of the two converging intervals is a mean between the ratio of the square roots of the spheres and the inverse ratio of the corresponding movements.

Moreover, from this you may understand that the ratio of the diverging movements is much greater than the ratio of the $\frac{3}{2}$th powers of the spheres, since the ratio of the $\frac{3}{2}$th powers is compounded with the squares of the ratio of the aphelial interval to the mean interval, and that of the mean to the perihelial.

4. In What Things Having to do with the Planetary Movements Have the Harmonic Consonances been Expressed by the Creator, and in What Way?

Accordingly, if the image of the retrogradation and stations is taken away and the proper movements of the planets in their real eccentric orbits are winnowed out, the following distinct things still remain in the planets: 1) The distances from the sun. 2) The periodic times. 3) The diurnal eccentric arcs. 4) The diurnal delays in those arcs. 5) The angles at the sun, and the diurnal arcs apparent to those as it were gazing from the sun. And again, all of these things, with the exception of the periodic times, are variable in the total circuit, most variable at the mean longitudes, but least at the extremes, when, turning away from one extreme longitude, they begin to return to the opposite. Hence when the planet is lowest and nearest to the sun and thereby delays the least in one degree of its eccentric, and conversely in one day traverses the greatest diurnal arc of its eccentric and appears fastest from the sun: then its movement remains for some time in this strength without preceptible variation, until, after passing the perihelion, the planet gradually begins to depart farther from the sun in a straight line; at that same time it delays longer in the degrees of its eccentric circle; or, if you consider the movement of one day, on the following day it goes forward less and appears even more slow from the sun until it has drawn close to the highest apsis and made its distance from the sun very great: for then longest of all does it delay in one degree of its eccentric; or on the contrary in one day it traverses its least arc and makes a much smaller apparent movement and the least of its total circuit.

Finally, all these things may be considered either as they exist in any one planet at different times or as they exist in different planets: whence, by the assumption of an infinite amount of time, all the affects of the circuit of one planet can concur in the same moment of time with all the affects of the circuit of another planet and be compared, and then the total eccentrics, as compared with one another, have the same ratio as their semidiameters or mean intervals; but the arcs of two eccentrics, which are similar or designated by the same number [of degrees], nevertheless have their true lengths unequal in the ratio of their eccentrics. For example, one degree in the sphere of Saturn is approximately twice as long as one degree in the sphere of Jupiter. And conversely, the diurnal arcs of the eccentrics, as expressed in astronomical terms, do not exhibit the ratio of the true journeys which the globes complete in one day [283] through the ether, because the single units in the wider circle of the upper planet denote a quarter part of the journey, but in the narrower circle of the lower planet a smaller part.

Therefore let us take the second of the things which we have posited, namely, the periodic times of the planets, which comprehend the sums made up of all the delays—long, middling, short—in all the degrees of the total circuit. And we found that from antiquity down to us, the planets complete their periodic returns around the sun, as follows in the table:

	Days	Minutes of a day	Therefore the mean diurnal movements		
			Min.	Sec.	Thirds
Saturn	10,759	12	2	0	27
Jupiter	4,332	37	4	59	8
Mars	686	59	31	26	31
Earth with Moon	365	15	59	8	11
Venus	224	42	96	7	39
Mercury	87	58	245	32	25

Accordingly, in these periodic times there are no harmonic ratios, as is easily apparent, if the greater periods are continuously halved, and the smaller are continuously doubled, so that, by neglecting the intervals of an octave, we can investigate the intervals which exist within one octave.

	Saturn	Jupiter	Mars	Earth	Venus	Mercury	
Halves	$10,759^D12'$ $5,379^D36'$ $2,689^D48'$ $1,344^D54'$ $672^D27'$	$4,332^D37'$ $2,166^D19'$ $1,083^D10'$ $541^D35'$	$686^D59'$	$365^D15'$	$224^D42'$ $449^D24'$	$87^D58'$ $175^D56'$ $351^D52'$	Doubles

All the last numbers, as you see, are counter to harmonic ratios and seem, as it were, irrational. For let 687, the number of days of Mars, receive as its measure 120, which is the number of the division of the chord: according to this measure Saturn will have 117 for one sixteenth of its period, Jupiter less than 95 for one eighth of its period, the earth less than 64, Venus more than 78 for twice its period, Mercury more than 61 for four times its period. These numbers do not make any harmonic ratio with 120, but their neighbouring numbers—60, 75, 80, and 96—do. And so, whereof Saturn has 120, Jupiter has approximately 97, the Earth more than 65, Venus more than 80, and Mercury less than 63. And whereof Jupiter has 120, the Earth has less than 81, Venus less than 100, Mercury less than 78. Likewise, whereof Venus has 120, the Earth has less than 98, Mercury more than 94. Finally, whereof the Earth has 120, Mercury has less than 116. But if the free choice of ratios had been effective here, consonances which are altogether perfect but not augmented or diminished would have been taken. Accordingly we find that God the Creator did not wish to introduce harmonic ratios between the sums of the delays added together to form the periodic times.

[284] And although it is a very probable conjecture (as relying on geometrical demonstrations and the doctrine concerning the causes of the planetary movements given in the *Commentaries on Mars*) that the bulks of the planetary bodies are in the ratio of the periodic times, so that the globe of Saturn is about thirty times greater than the globe of the Earth, Jupiter twelve times, Mars less than two, the Earth one and a half times greater than the globe of Venus and four times greater than the globe of Mercury: not therefore will even these ratios of bodies be harmonic.

But since God has established nothing without geometrical beauty, which was not bound by some other prior law of necessity, we easily infer that the periodic times have got their due lengths, and thereby the mobile bodies too have got their bulks, from something which is prior in the archetype, in order to express which thing these bulks and periods have been fashioned to this measure, as they seem disproportionate. But I have said that the periods are added up from the longest, the middling, and the slowest delays: accordingly geometrical fitnesses must be found either in these delays or in anything which may be prior to them in the mind of the Artisan. But the ratios of the delays are bound up with the ratios of the diurnal arcs, because the arcs have the inverse ratio of the delays. Again, we have said that the ratios of the delays and intervals of any one planet are the same. Then, as regards the single planets, there will be one and the same consideration of the following three: the arcs, the delays in equal arcs, and the distance of the arcs from the sun or the intervals. And because all these things are variable in the planets, there can be no doubt but that, if these things were allotted any geometrical beauty, then, by the sure design of the highest Artisan, they would have been received that at their extremes, at the aphelial and perihelial intervals, not at the mean intervals lying in between. For, given the ratios of the extreme intervals, there is no need of a plan to fit the intermediate ratios to a definite number. For they follow of themselves, by the necessity of planetary movement, from one extreme through all the intermediates to the other extreme.

Therefore the intervals are as follows, according to the very accurate observations of Tycho Brahe, by the method given in the *Commentaries on Mars* and investigated in very persevering study for seventeen years.

Intervals Compared with Harmonic Ratios[1]

Of Two Planets		Of Single Planets
Converging	*Diverging*	
		Saturn's aphelion 10,052. a. More than a minor whole tone $\dfrac{10,000}{9,000}$
$\dfrac{a}{d}=\dfrac{2}{1}$,	$\dfrac{b}{c}=\dfrac{5}{3}$	perihelion 8,968. b. Less than a major whole tone $\dfrac{10,000}{8,935}$
		Jupiter's aphelion 5,451. c. No concordant ratio but approximately 11 : 10, a discordant or diminished 6 : 5.
$\dfrac{c}{f}=\dfrac{4}{1}$,	$\dfrac{d}{e}=\dfrac{3}{1}$	perihelion 4,949. d.
		Mar's aphelion 1,665. e. Here 1662 : 1385 would be the consonance 6 : 5, and 1665 : 1332 would be 5 : 4
$\dfrac{e}{h}=\dfrac{5}{3}$,	$\dfrac{f}{g}=\dfrac{17}{20}$	perihelion 1,382. f.
		Earth's aphelion 1,018. g. Here 1025 : 984 would be the diesis 24 : 25. Therefore it does not have the diesis.
$\dfrac{g}{k}=\dfrac{2}{1\frac{1}{2}}$ viz. $\dfrac{1000}{710}, \dfrac{h}{i}=\dfrac{27}{20}$		perihelion 982. h.
		Venus' aphelion 729. i. Less than a sesquicomma.
$\dfrac{i}{m}=\dfrac{12}{5}$,	$\dfrac{k}{i}=\dfrac{243}{160}$	perihelion 719. k. More than one third of a diesis.
		Mercury's aphelion 470. l. 243 : 160, greater than a perfect fifth but less than a harmonic 8 : 5
		perihelion 307. m.

[285] Therefore the extreme intervals of no one planet come near consonances except those of Mars and Mercury.

But if you compare the extreme intervals of different planets with one another, some harmonic light begins to shine. For the extreme diverging intervals of Saturn and Jupiter make slightly more than the octave; and the con-

---

[1]GENERAL NOTE: Throughout this text Kepler's *concinna* and *inconcinna* are translated as "concordant" and "discordant." *Concinna* is usually used by Kepler of all intervals whose ratios occur within the "natural system" or the just intonation of the scale. *Inconcinna* refers to all ratios that lie outside of this system of tuning. "Consonant" (*consonans*) and "dissonant" (*dissonans*) refer to qualities which can be applied to intervals within the musical system, in other words to "concords." "Harmony" (*harmonia*) is used sometimes in the sense of "concordance" and sometimes in the sense of "consonance."

Genus durum and *genus molle* are translated either as "major mode" and "minor mode," or as "major scale" and "minor scale," or as "major kind" and "minor kind" (of consonances). The use of *modus*, to refer to the ecclesiastical modes, occurs only in Chapter 6.

As our present musical terms do not apply strictly to the music of the sixteenth and seventeenth centuries, a brief explanation of terms here may be useful. This material is taken from Kepler's *Harmonies of the World*, Book III.

An octave system in the minor scale (*Systema octavae in cantu molli*)

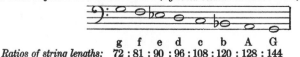

Ratios of string lengths: g   f   e   d   c   b   A   G
            72 : 81 : 90 : 96 : 108 : 120 : 128 : 144

verging, a mean between the major and minor sixths. So the diverging extremes of Jupiter and Mars embrace approximately the double octave; and the converging, approximately the fifth and the octave. But the diverging extremes of the Earth and Mars embrace somewhat more than the major sixth; the converging, an augmented fourth. In the next couple, the Earth and Venus, there is again the same augmented fourth between the converging extremes; but we lack any harmonic ratio between the diverging extremes: for it is less than the semi-octave (so to speak) *i.e.*, less than the square root of the ratio 2 : 1. Finally, between the diverging extremes of Venus and Mercury there is a ratio slightly

In the major scale (*In cantu duro*)

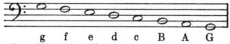

g   f   e   d   c   B   A   G

Ratios of string lengths: 360 : 405 : 432 : 480 : 540 : 576 : 640 : 720

As in all music, these scales can be repeated at one or more octaves above. The ratios would then all be halved, *i.e.*,

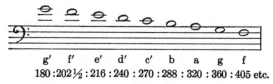

g'   f'   e'   d'   c'   b   a   g   f

180 : 202½ : 216 : 240 : 270 : 288 : 320 : 360 : 405 etc.

Various intervals which Kepler considers are:

80 : 81	*comma* (of Didymus), difference between major and minor whole tones ($\frac{8}{9} \div \frac{9}{10}$)	
24 : 25	*diesis* [difference between e – e flat or B – b flat or between a semitone and a minor whole tone ($\frac{15}{16} \div \frac{9}{10}$)]	
128 : 135	*lemma* [difference between a semitone and a major whole tone ($\frac{15}{16} \div \frac{8}{9}$)]	
243 : 256	*Plato's lemma* (not found in this system but in the Pythagorean tuning)	
15 : 16	*semitone*	minor mode between e flat – d, b flat – A major mode between e – d, B – A
9 : 10	*minor whole tone*	minor mode f – e flat, c – b flat major mode e – d, B – A
8 : 9	*major whole tone*	minor mode: g – f, d – c, A – G major mode: g – f, d – c, A – G
27 : 32	*sub-minor third* (major and minor modes: f – d, c – A)	
5 : 6	*minor third*	minor mode: e – flat – c, b flat – G major mode: g – e, d – B
4 : 5	*major third*	minor mode: g – e – flat, d – b – flat major mode: e – c, B – G
64 : 81	*ditone* (Pythagorean third) (major and minor modes: a – f)	
243 : 320	*lesser imperfect fourth* (inversion of "greater imperfect fifth") see below	
3 : 4	*perfect fourth*	minor mode: g – d, f – c, e flat – b flat, d – A, c – G major mode: g – d, f – c, e – B, , d – A, c – G
20 : 27	*greater imperfect fourth*	minor mode: b' flat – f major mode: a – e
32 : 45	*augmented fourth*	minor mode: a – e flat major mode: b – f
45 : 64	*diminished fifth*	minor mode: e – flat – A major mode: f – B
27 : 40	*lesser imperfect fifth*	minor mode: f – b flat major mode: e – A
2 : 3	*perfect fifth*	minor mode: g – c, d – G major mode: g – c, d – G

less than the octave compounded with the minor third; between the converging there is a slightly augmented fifth.

Accordingly, although one interval was somewhat removed from harmonic ratios, this success was an invitation to advance further. Now my reasonings were as follows: First, in so far as these intervals are lengths without movement, they are not fittingly examined for harmonic ratios, because movement is more properly the subject of consonances, by reason of speed and slowness. Second, inasmuch as these same intervals are the diameters of the spheres, it is believable that the ratio of the five regular solids applied proportionally is more dominant in them, because the ratio of the geometrical solid bodies to the celestial spheres (which are everywhere either encompassed by celestial matter, as the ancients hold, or to be encompassed successively by the accumulation of many revolutions) is the same as the ratio of the plane figures which may be inscribed in a circle (these figures engender the consonances) to the celestial circles of movements and the other regions wherein the movements take place. Therefore, if we are looking for consonances, we should look for them not in these

---

160 : 243	*greater imperfect fifth* (compound of ditone and minor third $^{64}\!/_{81} \times {}^5\!/_6$)	
81 : 128	*imperfect minor sixth* (minor and major modes: f – A)	
5 : 8	*minor sixth*	minor mode: e flat – G, b^{1b+} – d
		major mode: g – B, c′ – e
3 : 5	*major sixth*	minor mode: g – B flat, c′ – e flat
		major mode: e – G, b – d
64 : 27	*greater major sixth*	minor mode: d′ – f, a – c
		major mode: d′ – f, a – c
1 : 2	*octave* (g – G, a – A, b – B, b flat – b flat)	

All these are simple intervals. When one or more octaves are added to any simple intervals the resultant interval is a "compound" interval.

1 : 3   equals $\frac{1}{2} \times \frac{2}{3}$—an octave and a perfect fifth
3 : 32 equals $(\frac{1}{2})^3 \times \frac{3}{4}$—three octaves and a perfect fourth
1 : 20 equals $(\frac{1}{2})^4 \times {}^{16}\!/_{20}$—four octaves and a major third

Concords: All intervals from diesis downward on above list.
Consonances: Minor and major thirds and sixths, perfect fourth, fifth, and octave.
"Adulterine" consonances: sub-minor third, ditone, lesser imperfect fourth and fifth, greater imperfect fourth and fifth, imperfect minor sixth, greater major sixth.
Dissonances: All other intervals.

Throughout this work Kepler, after the fashion of the theorists of his time, uses the ratios of string lengths rather than the ratios of vibrations as is usually done today. String lengths are, of course, inversely proportionate to the vibrations. That is, string lengths 4 : 5 are expressed in vibrations as 5:4. This accounts for the descending order of the scale, which follows the increasing numerical order. It is an interesting fact that Kepler's minor and major scales are inversions of each other and hence, when expressed in ratios of vibrations, are in the opposite order from those in ratios of string lengths:

Notes resulting from ratios of vibrations

72 : 81 : 90 : 96 : 108 : 120 : 128 : 144   360 : 405 : 432 : 480 : 540 : 576 : 640 : 720

Notes resulting from ratios of string lengths

An arbitrary pitch G is chosen to situate these ratios. This g or "gamma" was usually the lowest tone of the sixteenth-century musical gamut.   ELLIOTT CARTER, JR.

intervals in so far as they are the semidiameters of spheres but in them in so far as they are the measures of the movements, *i.e.*, in the movements themselves, rather. Absolutely no other than the mean intervals can be taken as the semidiameters of the spheres; but we are here dealing with the extreme intervals. Accordingly, we are not dealing with the intervals in respect to their spheres but in respect to their movements.

Accordingly, although for these reasons I had passed on to the comparison of the extreme movements, at first the ratios of the movements remained the same in magnitude as those which were previously the ratios of the intervals, only inverted. Wherefore too, certain ratios, which are discordant and foreign to harmonies, as before, have been found between the movements. But once again I judged that this happened to me deservedly, because I compared with one another eccentric arcs which are not expressed and numbered by a measure of the same magnitude but are numbered in degrees and minutes which are of diverse magnitude in diverse planets, nor do they from our place give the appearance of being as great as the number of each says, except only at the centre of the eccentric of each planet, which centre rests upon no body; and hence it is also unbelievable that there is any sense or natural instinct in that place in the world which is capable of perceiving this; or, rather, it was impossible, if I was comparing the eccentric arcs of different planets with respect to their appearance at their centres, which are different for different planets. But if diverse apparent magnitudes are compared with one another, they ought to be apparent in one place in the world in such a way that that which possesses the faculty of comparing them may be present in that place from which they are all apparent. Accordingly, I judged that the appearance of these eccentric arcs should be removed from the mind or else should be formed differently. But if I removed the appearance and applied my mind to the diurnal journeys of the planets, I saw that I had to employ the rule which I gave in Article ix of the preceding chapter. [286] Accordingly if the diurnal arcs of the eccentric are multiplied by the mean intervals of the spheres, the following journeys are produced:

	Diurnal movements	*Mean intervals*	*Diurnal journeys*
Saturn at aphelion	1'53"	9510	1065
at perihelion	2'7"		1208
Jupiter at aphelion	4'44"	5200	1477
at perihelion	5'15"		1638
Mars at aphelion	28'44"	1524	2627
at perihelion	34'34"		3161
Earth at aphelion	58'6"	1000	3486
at perihelion	60'13"		3613
Venus at aphelion	95'29"	724	4149
at perihelion	96'50"		4207
Mercury at aphelion	201'0"	388	4680
at perihelion	307'3"		7148

Thus Saturn traverses barely one seventh of the journey of Mercury; and hence, as Aristotle judged consonant with reason in Book ii of *On the Heavens*, the planet which is nearer the sun always traverses a greater space than the planet which is farther away—as cannot hold in the ancient astronomy.

And indeed, if we weigh the thing fairly carefully, it will appear to be not very probable that the most wise Creator should have established harmonies between the planetary journeys in especial. For if the ratios of the journeys are harmonic, all the other affects which the planets have will be necessitated and bound up with the journeys, so that there is no room elsewhere for establishing harmonies. But whose good will it be to have harmonies between the journeys, or who will perceive these harmonies? For there are two things which disclose to us harmonies in natural things: either light or sound: light apprehended through the eyes or hidden senses proportioned to the eyes, and sound through the ears. The mind seizes upon these forms and, whether by instinct (on which Book IV speaks profusely) or by astronomical or harmonic ratiocination, discerns the concordant from the discordant. Now there are no sounds in the heavens, nor is the movement so turbulent that any noise is made by the rubbing against the ether. Light remains. If light has to teach these things about the planetary journeys, it will teach either the eyes or a sensorium analogous to the eyes and situated in a definite place; and it seems that sense-perception must be present there in order that light of itself may immediately teach. Therefore there will be sense-perception in the total world, namely in order that the movements of all the planets may be presented to sense-perceptions at the same time. For that former route—from observations through the longest detours of geometry and arithmetic, through the ratios of the spheres and the other things which must be learned first, down to the journeys which have been exhibited—is too long for any natural instinct, for the sake of moving which it seems reasonable that the harmonies have been introduced.

Therefore with everything reduced to one view, I concluded rightly [287] that the true journeys of the planets through the ether should be dismissed, and that we should turn our eyes to the apparent diurnal arcs, according as they are all apparent from one definite and marked place in the world—namely, from the solar body itself, the source of movement of all the planets; and we must see, not how far away from the sun any one of the planets is, nor how much space it traverses in one day (for that is something for ratiocination and astronomy, not for instinct), but how great an angle the diurnal movement of each planet subtends in the solar body, or how great an arc it seems to traverse in one common circle described around the sun, such as the ecliptic ,in order that these appearances, which were conveyed to the solar body by virtue of light, may be able to flow, together with the light, in a straight line into creatures, which are partakers of this instinct, as in Book IV we said the figure of the heavens flowed into the foetus by virtue of the rays.

Therefore, if you remove from the proper planetary movement the parallaxes of the annual orbit, which gives them the mere appearances of stations and retrogradations, Tycho's astronomy teaches that the diurnal movements of the planets in their orbits (which are apparent as it were to spectator at the sun) are as shown in the table on the opposite page.

Note that the great eccentricity of Mercury makes the ratio of the movements differ somewhat from the ratio of the square of the distances. For if you make the square of the ratio of 100, the mean distance, to 121, the aphelial distance, be the ratio of the aphelial movement to the mean movement of 245′32″, then an aphelial movement of 167 will be produced; and if the square of the ratio of 100 to 79, the perihelial distance, be the ratio of the perihelial to the same mean movement, then the perihelial movement will become 393; and both cases are

Harmonies Between Two Planets		Apparent Diurnal Movements		Harmonies Between the Movements of Single Planets
Diverging	*Converging*			
		Saturn at aphelion	1'46'' a.	1 : 48'' : 2'15''=4 : 5,
		at perihelion	2'15'' b.	major third
$\dfrac{a}{d}=\dfrac{1}{3}$,	$\dfrac{b}{c}=\dfrac{1}{2}$			
		Jupiter at aphelion	4'30'' c.	4'35'' : 5'30''=5 : 6,
		at perihelion	5'30'' d.	minor third
$\dfrac{c}{f}=\dfrac{1}{8}$,	$\dfrac{d}{e}=\dfrac{5}{24}$			
		Mars at aphelion	26'14'' e.	25'21'' : 38'1''=2 : 3,
		at perihelion	38'1'' f.	the fifth
$\dfrac{e}{h}=\dfrac{5}{12}$,	$\dfrac{f}{g}=\dfrac{2}{3}$			
		Earth at aphelion	57'3'' g.	57'28'' : 61'18''=15 : 16,
		at perihelion	61'18'' h.	semitone
$\dfrac{g}{k}=\dfrac{3}{5}$,	$\dfrac{h}{i}=\dfrac{5}{8}$			
		Venus at aphelion	94'50'' i.	94'50'' : 98'47''=24 : 25,
		at perihelion	97'37'' k.	diesis
$\dfrac{i}{m}=\dfrac{1}{4}$,	$\dfrac{k}{l}=\dfrac{3}{5}$			
		Mercury at aphelion 164'0'' l.		164'0'' : 394'0''=5 : 12,
		at perihelion 384'0'' m.		octave and minor third

greater than I have here laid down, because the mean movement at the mean anomaly, viewed very obliquely, does not appear as great, *viz.*, not as great as 245'32'', but about 5' less. Therefore, too, lesser aphelial and perihelial movements will be elicited. But the aphelial [appears] lesser and the perihelial greater, on account of theorem 8, Euclid's *Optics*, as I remarked in the preceding Chapter, Article vi.

Accordingly, I could mentally presume, even from the ratios of the diurnal eccentric arcs given above, that there were harmonies and concordant intervals between these extreme apparent movements of the single planets, since I saw that everywhere there the square roots of harmonic ratios were dominant, but knew that the ratio of the apparent movements was the square of the ratio of the eccentric movements. But it is possible by experience itself, or without any ratiocination to prove what is affirmed, as you see [288] in the preceding table. The ratios of the apparent movements of the single planets approach very close to harmonies, in such fashion that Saturn and Jupiter embrace slightly more than the major and minor thirds, Saturn with a ratio of excess of 53 : 54, and Jupiter with one of 54 : 55 or less, namely approximately a sesquicomma; the Earth, slightly more (namely 137 : 138, or barely a semicomma) than a semitone; Mars somewhat less (namely 29 : 30, which approaches 34 : 35 or 35 : 36) than a fifth; Mercury exceeds the octave by a minor third rather than a whole tone, *viz.*, it is about 38 : 39 (which is about two commas, *viz.*, 34 : 35 or 35 : 36) less than a whole tone. Venus alone falls short of any of the concords the diesis; for its ratio is between two and three commas, and it exceeds two thirds of a diesis, and is about 34 : 35 or 35 : 36, a diesis diminished by a comma.

The moon, too, comes into this consideration. For we find that its hourly apogeal movement in the quadratures, *viz.*, the slowest of all its movements, to be 26'26"; its perigeal movement in the syzygies, *viz.*, the fastest of all, 35'12", in which way the perfect fourth is formed very precisely. For one third of 26'26" is 8'49", the quadruple of which is 35'16". And note that the consonance of the perfect fourth is found nowhere else between the apparent movements; note also the analogy between the fourth in consonances and the quarter in the phases. And so the above things are found in the movements of the single planets.

But in the extreme movements of two planets compared with one another, the radiant sun of celestial harmonies immediately shines at first glance, whether you compare the diverging extreme movements or the converging. For the ratio between the diverging movements of Saturn and Jupiter is exactly the duple or octave; that between the diverging, slightly more than triple or the octave and the fifth. For one third of 5'30" is 1'50", although Saturn has 1'46" instead of that. Accordingly, the planetary movements will differ from a consonance by a diesis more or less, *viz.*, 26 : 27 or 27 : 28; and with less than one second acceding at Saturn's aphelion, the excess will be 34 : 35, as great as the ratio of the extreme movements of Venus. The diverging and converging movements of Jupiter and Mars are under the sway of the triple octave and the double octave and a third, but not perfectly. For one eighth of 38'1" is 4'45", although Jupiter has 4'30"; and between these numbers there is still a difference of 18 : 19, which is a mean between the semitone of 15 : 16 and the diesis of 24 : 25, namely, approximately a perfect lemma of 128 : 135.[1] Thus one fifth of 26'14" is 5'15", although Jupiter has 5'30"; accordingly in this case the quintuple ratio is diminished in the ratio of 21 : 22, the augment in the case of the other ratio, *viz.*, approximately a diesis of 24 : 25.

The consonance 5 : 24 comes nearer, which compounds a minor instead of a major third with the double octave. For one fifth of 5'30" is 1'6", which if multiplied by 24 makes 26'24", does not differ by more than a semicomma. Mars and the Earth have been allotted the least ratio, exactly the sesquialteral or perfect fifth: for one third of 57'3" is 19'1", the double of which is 38'2", which is Mars' very number, *viz.*, 38'11". They have also been allotted the greater ratio of 5 : 12, the octave and minor third, but more imperfectly. For one twelfth of 61'18" is 5'6½", which if multiplied by 5 gives 25'33", although instead of that Mars has 26'14". Accordingly, there is a deficiency of a diminished diesis approximately, *viz.*, 35 : 36. But the Earth and Venus together have been allotted 3 : 5 as their greatest consonance and 5 : 8 as their least, the major and minor sixths, but again not perfectly. For one fifth of 97'37", which if multiplied by 3 gives 58'33", which is greater than the movement of the Earth in the ratio 34 : 35, which is approximately 35 : 36: by so much do the planetary ratios differ from the harmonic. Thus one eighth of 94'50" is 11'51"+, five times which is 59'16", which is approximately equal to the mean movement of the Earth. Wherefore here the planetary ratio is less than the harmonic [289] in the ratio of 29 : 30 or 30 : 31, which is again approximately 35 : 36, the diminished diesis; and thereby this least ratio of these planets approaches the consonance of the perfect fifth. For one third of 94'50" is 31'37", the double of which is 63'14", of which the 61'18" of the perihelial movement of the Earth falls short in the ratio

[1]*cf.* Footnote to *Intervals Compared with Harmonic Ratios*, p. 186.

of 31 : 32, so that the planetary ratio is exactly a mean between the neighbouring harmonic ratios. Finally, Venus and Mercury have been allotted the double octave as their greatest ratio and the major sixth as their least, but not absolute-perfectly. For one fourth of 384′ is 96′0″, although Venus has 94′50″. Therefore the quadruple adds approximately one comma. Thus one fifth of 164′ is 32′48″, which if multiplied by 3 gives 98′24″, although Venus has 97′37″. Therefore the planetary ratio is diminished by about tow thirds of a comma, *i.e.*, 126 : 127.

Accordingly the above consonances have been ascribed to the planets; nor is there any ratio from among the principal comparisons (*viz.*, of the converging and diverging extreme movements) which does not approach so nearly to some consonance that, if strings were tuned in that ratio, the ears would not easily discern their imperfection—with the exception of that one excess between Jupiter and Mars.

Moreover, it follows that we shall not stray far away from consonances if we compare the movements of the same field. For if Saturn's 4 : 5 comp. 53 : 54 are compounded with the intermediate 1 : 2, the product is 2 : 5 comp. 53 : 54, which exists between the aphelial movements of Saturn and Jupiter. Compound with that Jupiter's 5 : 6 comp. 54 : 55, and the product is 5 : 12 comp 54 : 55, which exist between the perihelial movements of Saturn and Jupiter. Thus compound Jupiter's 5 : 6 comp. 54 : 55 with the intermediate ensuing ratio of 5 : 24 comp. 158 : 157, the product will be 1 : 6 comp. 36 : 35 between the aphelial movements. Compound the same 5 : 24 comp. 158 : 157 with Mars' 2 : 3 comp. 30 : 29, and the product will be 5 : 36 comp. 25 : 24 approximately, *i.e.*, 125 : 864 or about 1 : 7, between the perihelial movements. This ratio is still alone discordant. With 2 : 3 the third ratio among the intermediates, compound Mars' 2 : 3 less 29 : 30; the result will be 4 : 9 comp. 30 : 29, *i.e.*, 40 : 87, another discord between the aphelial movements. If instead of Mars' you compound the Earth's 15 : 16 comp. 137 : 138, you will make 5 : 8 comp. 137 : 138 between the perihelial movements. And if with the fourth of the intermediates, 5 : 8 comp. 31 : 30, or 2 : 3 comp. 31 : 32, you compound the Earth's 15 : 16 comp. 137 : 138, the product will be approximately 3 : 5 between the aphelial movements of the Earth and Venus. For one fifth of 94′50″ is 18′58″, the triple of which is 56′54″, although the Earth has 57′3″. If you compound Venus' 34 : 35 with the same ratio, the result will be 5 : 8 between the perihelial movements. For one eighth of 97′37″ is 12′12″+ which if multiplied by 5 gives 61′1″, although the Earth has 61′18″. Finally, if with the last of the intermediate ratios, 3 : 5 comp. 126 : 127 you compound Venus' 34 : 35, the result is 3 : 5 comp. 24 : 25, and the interval, compounded of both, between the aphelial movements, is dissonant. But if you compound Mercury's 5 : 12 comp. 38 : 39, the double octave or 1 : 4 will be diminished by approximately a whole diesis, in proportion to the perihelial movements.

Accordingly, perfect consonances are found: between the converging movements of Saturn and Jupiter, the octave; between the converging movements of Jupiter and Mars, the octave and minor third approximately; between the converging movements of Mars and the Earth, the fifth; between their perihelial, the minor sixth; between the extreme converging movements of Venus and Mercury, the major sixth; between the diverging or even between the perihelial, the double octave: whence without any loss to an astronomy which has been built, most subtly of all, upon Brahe's observations, it seems that the residual very slight discrepancy can be discounted, especially in the movements of Venus and Mercury.

But you will note that where there is no perfect major consonance, as between Jupiter and Mars, there alone have I found the placing of the solid figure to be approximately perfect, since the perihelial distance of Jupiter is approximately three times the aphelial distance of Mars, in such fashion that this pair of planets strives after the perfect consonance in the intervals which it does not have in the movements.

[290] You will note, furthermore, that the major planetary ratio of Saturn and Jupiter exceeds the harmonic, *viz.*, the triple, by approximately the same quantity as belongs to Venus; and the common major ratio of the converging and diverging movements of Mars and the Earth are diminished by approximately the same. You will note thirdly that, roughly speaking, in the upper planets the consonances are established between the converging movements, but in the lower planets, between movements in the same field. And note fourthly that between the aphelial movements of Saturn and the Earth there are approximately five octaves; for one thirty-second of 57'3" is 1'47", although the aphelial movement of Saturn is 1'46".

Furthermore, a great distinction exists between the consonances of the single planets which have been unfolded and the consonances of the planets in pairs. For the former cannot exist at the same moment of time, while the latter absolutely can; because the same planet, moving at its aphelion, cannot be at the same time at the opposite perihelion too, but of two planets one can be at its aphelion and the other at its perihelion at the same moment of time. And so the ratio of plain-song or monody, which we call choral music and which alone was known to the ancients,[1] to polyphony—called "figured song,";[2] the invention of the latest generations—is the same as the ratio of the consonances which the single planets designate to the consonances of the planets taken together. And so, further on, in Chapters 5 and 6, the single planets will be compared to the choral music of the ancients and its properties will be exhibited in the planetary movements. But in the following chapters, the planets taken together and the figured modern music will be shown to do similar things.

5. IN THE RATIOS OF THE PLANETARY MOVEMENTS WHICH ARE APPARENT AS IT WERE TO SPECTATORS AT THE SUN, HAVE BEEN EXPRESSED THE PITCHES OF THE SYSTEM, OR NOTES OF THE MUSICAL SCALE, AND THE MODES OF SONG [GENERA CANTUS], THE MAJOR AND THE MINOR[3]

Therefore by now I have proved by means of numbers gotten on one side from astronomy and on the other side from harmonics that, taken in every which way, harmonic ratios hold between these twelve termini or movements of the six planets revolving around the sun or that they approximate such ratios within an imperceptible part of least concord. But just as in Book III in the first chapter, we first built up the single harmonic consonances separately, and then

---

[1]The choral music of the Greeks was monolinear, everyone singing the same melody together.—E. C., Jr.

[2]In plain-song all the time values of the notes were approximately equal, while in "figured song" time values of different lengths were indicated by the notes, which gave composers an opportunity both to regulate the way different contrapuntal parts joined together and to produce many expressive effects. Practically all melodies since this time are in "figured song" style.—E. C., Jr.

[3]See note to *Intervals Compared with Harmonic Ratios*, p. 186.

we joined together all the consonances—as many as there were—in one common system or musical scale, or, rather, in one octave of them which embraces the rest in power, and by means of them we separated the others into their degrees or pitches [*loca*] and we did this in such a way that there would be a scale; so now also, after the discovery of the consonances [*harmoniis*] which God Himself has embodied in the world, we must consequently see whether those single consonances stand so separate that they have no kinship with the rest, or whether all are in concord with one another. Notwithstanding it is easy to conclude, without any further inquiry, that those consonances were fitted together by the highest prudence in such fashion that they move one another about within one frame, so to speak, and do not jolt one another out of it; since indeed we see that in such a manifold comparison of the same terms there is no place where consonances do not occur. For unless in one scale all the consonances were fitted to all, it could easily have come about (and it has come about wherever necessity thus urges it) that many dissonances should exist. For example, if someone had set up a major sixth between the first and the second term, and likewise a major third between the second and the third term, without taking the first into account, then he would admit a dissonance and the discordant interval 12 : 25 be-between the first and third.

But come now, let us see whether that which we have already inferred by reasoning is really found in this way. [291] But let me premise some cautions, that we may be the less impeded in our progress. First, for the present, we must conceal those augments or diminutions which are less than a semitone; for we shall see later on what causes they have. Second, by continuous doubling or contrary halving of the movements, we shall bring everything within the range of one octave, on account of the sameness of consonance in all the octaves.

Accordingly the numbers wherein all the pitches or clefs [*loca seu claves*] of the octave system are expressed have been set out in a table in Book III, Chapter 7[1],

[1]The table is as follows:

Concordant Intervals	Lengths of Strings	In familiar notes	
	1080	High g	
Semitone			
	1152	f ♯	
Lemma			
	1215	f	
Semitone			
	1296	e	
Diesis			
	1350	e ♭	
Semitone			
	1440	d	
Semitone			
	1536	c ♯	
Lemma			
	1620	c	
Semitone			
	1728	b	
Diesis			
	1800	b ♭	
Semitone			
	1920	A	
Semitone			
	2048	G ♯	
Lemma			
	2160	Low G	

i.e., understand these numbers of the length of two strings. As a consequence, the speeds of the movements will be in the inverse ratios.

Now let the planetary movements be compared in terms of parts continuously halved. Therefore

Movement of Mercury at perihelion,	7th subduple, or $\frac{1}{128}$,	3'0"
at aphelion,	6th subduple, or $\frac{1}{64}$,	2'34"
Movement of Venus at perihelion,	5th subduple, or $\frac{1}{32}$,	3'3"
at aphelion,	5th subduple, or $\frac{1}{32}$,	2'58"
Movement of Earth at perihelion,	5th subduple, or $\frac{1}{32}$,	1'55"
at aphelion,	5th subduple, or $\frac{1}{32}$,	1'47"
Movement of Mars at perihelion,	4th subduple, or $\frac{1}{16}$,	2'23"
at aphelion,	3rd subduple, or $\frac{1}{8}$,	3'17"
Movement of Jupiter at perihelion,	subduple, or $\frac{1}{2}$,	2'45"
at aphelion,	subduple, or $\frac{1}{2}$,	2'15"
Movement of Saturn at perihelion,		2'15"
at aphelion,		1'46"

Now the aphelial movement of Saturn at its slowest—*i.e.*, the slowest movement—marks *G*, the lowest pitch in the system with the number 1'46". Therefore the aphelial movement of the Earth will mark the same pitch, but five octaves higher, because its number is 1'47", and who wants to quarrel about one second in the aphelial movement of Saturn? But let us take it into account, nevertheless; the difference will not be greater than 106 : 107, which is less than a comma. If you add 27", one quarter of this 1'47", the sum will be 2'14", although the perihelial movement of Saturn has 2'15"; similarly the aphelial movement of Jupiter, but one octave higher. Accordingly, these two movements mark the note *b*, or else are very slightly higher. Take 36", one third of 1'47", and add it to the whole; you will get as a sum 2'23" for the note *c*; and here's the perihelion of Mars of the same magnitude but four octaves higher. To this same 1'47" add also 54", half of it, and the sum will be 2'41" for the note *d*; and here the perihelion of Jupiter is at hand, but one octave higher, for it occupies the nearest number, *viz.*, 2'45". If you add two thirds, *viz.*, 1'11", the sum will be 2'58"; and here's the aphelion of Venus at 2'58". Accordingly, it will mark the pitch or the note *e*, but five octaves higher. And the perihelial movement of Mercury, which is 3'0", does not exceed it by much but is seven octaves higher. Finally, divide the double of 1'47", *viz.*, 3'34", into nine parts and subtract one part of 24" from the whole; 3'10" will be left for the note *f*, which the 3'17" of the aphelial movement of Mars marks approximately but three octaves higher; and this number is slightly greater than the just number and approaches the note *f* sharp. For if one sixteenth of 3'34", *viz.*, 13½", is subtracted from 3'34", then 3'20½" is left, to which 3'17" is very near. And indeed in music *f* sharp is often employed in place of *f*, as we can see everywhere.

Accordingly all the notes of the major scale [*cantus duri*] (except the note *a* which was not marked by harmonic division, in Book III, Chapter 2) are marked by all the extreme movements of the planets, except the perihelial movements of Venus and the Earth [292] and the aphelial movement of Mercury, whose number, 2'34", approaches the note *c* sharp. For subtract from the 2'41" of *d* one sixteenth or 10", and 2'30" remains for the note *c* sharp. Thus only the perihelial movement of Venus and the Earth are missing from this scale, as you may see in the table.

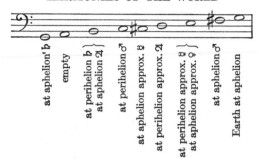

On the other hand, if the beginning of the scale is made at 2′15″, the aphelial movement of Saturn, and we must express the note *G* in those degrees: then for the note *A* is 2′32″, which closely approaches the aphelial movement of Mercury; for the note *b* flat, 2′42″, which is approximately the perihelial movement of Jupiter, by the equipollence of octaves; for the note *c*, 3′0″, approximately the perihelial movement of Mercury and Venus; for the note *d*, 3′23″ and the aphelial movement of Mars is not much graver, *viz.*, 3′17″, so that here the number is about as much less than its note as previously the same number was greater than its note; for the note *e* flat, 3′36″, which the aphelial movement of the Earth approximates; for the note *e*, 3′50″, and the perihelial movement of the Earth is 3′49″; but the aphelial movement of Jupiter again occupies *g*. In this way, all the notes except *f* are expressed within one octave of the minor scale by most of the aphelial and perihelial movements of the planets, especially by those which were previously omitted, as you see in the table.

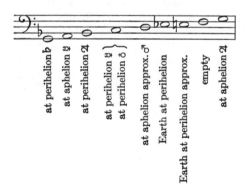

Previously, however, *f* sharp was marked and *a* omitted; now *a* is marked, *f* sharp is omitted; for the harmonic division in Chapter 2 also omitted the note *f*.

Accordingly, the musical scale or system of one octave with all its pitches, by means of which natural song[1] is transposed in music, has been expressed in the heavens by a twofold way and in two as it were modes of song. There is this sole difference: in our harmonic sectionings both ways start together from one and the same terminus G; but here, in the planetary movements, that which was previously *b* now becomes *G* in the minor mode.

[1]Natural song: music in the basic major or minor system without accidentals. E. C., Jr.

In the celestial movements, as follows:

By harmonic sectionings, as follows:

For as in music 2160 : 1800, or 6 : 5, so in that system which the heavens express, 1728 : 1440, namely, also 6 : 5; and so for most of the remaining, 2160 : 1800, 1620, 1440, 1350, 1080 as 1728 : 1440, 1296, 1152, 1080, 864.

Accordingly you won't wonder any more that a very excellent order of sounds or pitches in a musical system or scale has been set up by men, since you see that they are doing nothing else in this business except to play the apes of God the Creator and to act out, as it were, a certain drama of the ordination of the celestial movements.

But there still remains another way whereby we may understand the twofold musical scale in the heavens, where one and the same system but a twofold tuning [*tensio*] is embraced, one at the aphelial movement of Venus, the other at the perihelial, because the variety of movements of this planet is of the least magnitude, as being such as is comprehended within the magnitude of the diesis, the least concord. And the aphelial tuning [*tensio*], as above, has been given to the aphelial movements of Saturn, the Earth, Venus, and (relatively speaking) Jupiter, in *G*, *e*, *b*, but to the perihelial movements of Mars and (relatively speaking) Saturn and, as is apparent at first glance, to those of Mercury, in *c*, *e*, and *b*. On the other hand, the perihelial tuning supplies a pitch even for the aphelial movements of Mars, Mercury, and (relatively speaking) Jupiter, but to the perihelial movements of Jupiter, Venus, and (relatively speaking) Saturn, and to a certain extent to that of the Earth and indubitably to that of Mercury too. For let us suppose that now not the aphelial movement of Venus but the 3′3″ of the perihelial gets the pitch of *e*; it is approached very closely by the 3′0″ of the perihelial movement of Mercury, through a double octave, at the end of Chapter 4. But if 18″ or one tenth of this perihelial movement of Venus is subtracted, 2′45″ remains, the perihelion of Jupiter, which occupies the pitch of *d*; and if one fifteenth or 12″ is added, the sum will be 3′15″, approximately the perihelion of Mars which occupies the pitch of *f*; and thus in *b*, the perihelial movement of Saturn and the aphelial movement of Jupiter have approximately the same tuning. But one eighth, or 23″, if multiplied by 5, gives 1′55″, which is the perihelial movement of the Earth; and, although it does not square with the foregoing in the same scale, as it does not give the interval 5 : 8 below *e* nor 24 : 25 above *G*, nevertheless if now the perihelial movement of Venus and so too the aphelial movement of Mercury, outside of the order, occupy the pitch *e*-flat instead of *e*, then there the perihelial movement of the Earth will occupy the pitch of *G*, and the aphelial movement of Mercury is in concord, because 1′1″, or one third of 3′3″, if multiplied by 5, gives 5′5″, half of which, or

2′32″, approximates the aphelion of Mercury, which in this extraordinary adjustment will occupy the pitch of *c*. Therefore, all these movements are of the same tuning with respect to one another; but the perihelial movement of Venus together with the three (or five) prior movements, *viz.*, in the same harmonic mode, divides the scale differently from the aphelial movement of the same in its tuning, *viz.*, in the major mode [*denere duro*]. Moreover, the perihelial movement of Venus, together with the two posterior movements, divides the same scale differently, *viz.*, not into concords but merely into a different order of concords, namely one which belongs to the minor mode [*generis mollis*].

But it is sufficient to have laid before the eyes in this chapter what is the case casually, but it will be disclosed in Chapter 9 by the most lucid demonstrations why each and every one of these things was made in this fashion and what the causes were not merely of harmony but even of the very least discord.

6. IN THE EXTREME PLANETARY MOVEMENTS THE MUSICAL MODES OR TONES HAVE SOMEHOW BEEN EXPRESSED

[294] This follows from the aforesaid and there is no need of many words; for the single planets somehow mark the pitches of the system with their perihelial movement, in so far as it has been appointed to the single planets to traverse a certain fixed interval in the musical scale comprehended by the definite notes of it or the pitches of the system, and beginning at that note or pitch of each planet which in the preceding chapter fell to the aphelial movement of that planet: *G* to Saturn and the Earth, *b* to Jupiter, which can be transposed higher to *G*, *f*-sharp to Mars, *e* to Venus, *a* to Mercury in the higher octave. See the single movements in the familiar terms of notes. They do not form articulately the intermediate positions, which you here see filled by notes, as they do the extremes, because they struggle from one extreme to the opposite not by leaps and intervals but by a continuum of tunings and actually traverse all the means (which are potentially infinite)—which cannot be expressed by me in any other way than by a continuous series of intermediate notes. Venus remains approximately in unison and does not equal even the least of the concordant intervals in the difference of its tension.

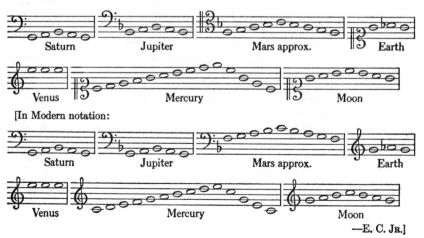

[In Modern notation:

Saturn   Jupiter   Mars approx.   Earth

Venus   Mercury   Moon

—E. C. JR.]

But the signature of two accidentals (flats) in a common staff and the formation of the skeletal outline of the octave by the inclusion of a definite concordant interval are a certain first beginning of the distinction of Tones or Modes [*modorum*]. Therefore the musical Modes have been distributed among the planets. But I know that for the formation and determination of distinct Modes many things are requisite, which belong to human song, as containing (a) distinct [order of] intervals; and so I have used the word *somehow*.

But the harmonist will be free to choose his opinion as to which Mode each planet expresses as its own, since the extremes have been assigned to it here. From among the familiar Modes, I should give to Saturn the Seventh or Eighth, because if you place its key-note at G, the perihelial movement ascends to b; to Jupiter, the First or Second Mode, because its aphelial movement has been fitted to G and its perihelial movement arrives at b flat; to Mars, the Fifth or Sixth Mode, not only because Mars comprehends approximately the perfect fifth, which interval is common to all the Modes, but principally because when it is reduced with the others to a common system, it attains c with its perihelial movement and touches f with its aphelial, which is the key-note of the Fifth or Sixth Mode or Tone; I should give the Third or Fourth Mode to the Earth, because its movement revolves within a semitone, while the first interval of those Modes is a semitone; but to Mercury will belong indifferently all the Modes or Tones on account of the greatness of its range; to Venus, clearly none on account of the smallness of its range; but on account of the common system the Third and Fourth Mode, because with reference to the other planets it occupies e. (The Earth sings MI, FA, MI so that you may infer even from the syllables that in this our domicile MIsery and FAmine obtain.)[1]

7. The Universal Consonances of All Six Planets, Like Common Four-Part Counterpoint, Can Exist

[295] But now, Urania, there is need for louder sound while I climb along the harmonic scale of the celestial movements to higher things where the true archetype of the fabric of the world is kept hidden. Follow after, ye modern musicians, and judge the thing according to your arts, which were unknown to antiquity. Nature, which is never not lavish of herself, after a lying-in of two thousand years, has finally brought you forth in these last generations, the first true images of the universe. By means of your concords of various voices, and through your ears, she has whispered to the human mind, the favorite daughter of God the Creator, how she exists in the innermost bosom.

(Shall I have committed a crime if I ask the single composers of this generation for some artistic motet instead of this epigraph? The Royal Psalter and the other Holy Books can supply a text suited for this. But alas for you! No more than six are in concord in the heavens. For the moon sings here monody separately, like a dog sitting on the Earth. Compose the melody; I, in order that the book may progress, promise that I will watch carefully over the six parts. To him who more properly expresses the celestial music described in this work, Clio will give a garland, and Urania will betroth Venus his bride.)

It has been unfolded above what harmonic ratios two neighbouring planets would embrace in their extreme movements. But it happens very rarely that two, especially the slowest, arrive at their extreme intervals at the same time; For example, the apsides of Saturn and Jupiter are about 81° apart. According-

[1]See note on hexachordal system.

ly, while this distance between them measures out the whole zodiac by definite twenty-year leaps[1], eight hundred years pass by, and nonetheless the leap which concludes the eighth century, does not carry precisely to the very apsides; and if it digresses much further, another eight hundred years must be awaited, that a more fortunate leap than that one may be sought; and the whole route must be repeated as many times as the measure of digression is contained in the length of one leap. Moreover, the other single pairs of planets have periods like that, although not so long. But meanwhile there occur also other consonances of two planets, between movements whereof not both are extremes but one or both are intermediate; and those consonances exist as it were in different tunings [tensionibus]. For, because Saturn tends from G to b, and slightly further, and Jupiter from b to d and further; therefore between Jupiter and Saturn there can exist the following consonances, over and above the octave: the major and minor third and the perfect fourth, either one of the thirds through the tuning which maintains the amplitude of the remaining one, but the perfect fourth through the amplitude of a major whole tone. For there will be a perfect fourth not merely from G of Saturn to cc of Jupiter but also from A of Saturn to dd of Jupiter and through all the intermediates between the G and A of Saturn and the cc and dd of Jupiter. But the octave and the perfect fifth exist solely at the points of the apsides. But Mars, which got a greater interval as its own, received it in order that it should also make an octave with the upper planets through some amplitude of tuning. Mercury received an interval great enough for it to set up almost all the consonances with all the planets within one of its periods, which is not longer than the space of three months. On the other hand, the Earth, and Venus much more so, on account of the smallness of their intervals, limit the consonances, which they form not merely with the others but with one another in especial, to visible fewness. But if three planets are to concord in one harmony, many periodic returns are to be awaited; nevertheless there are many consonances, so that they may so much the more easily take place, while each nearest consonance follows after its neighbour, and very often threefold consonances are seen to exist between Mars, the Earth, and Mercury. But the consonances of four planets now begin to be scattered throughout centuries, and those of five planets throughout thousands of years.

But that all six should be in concord [296] has been fenced about by the longest intervals of time; and I do not know whether it is absolutely impossible for this to occur twice by precise evolving or whether that points to a certain beginning of time, from which every age of the world has flowed.

But if only one sextuple harmony can occur, or only one notable one among many, indubitably that could be taken as a sign of the Creation. Therefore we must ask, in exactly how many forms are the movements of all six planets reduced to one common harmony? The method of inquiry is as follows: let us begin with the Earth and Venus, because these two planets do not make more than two consonances and (wherein the cause of this thing is comprehended) by means of very short intensifications of the movements.

Therefore let us set up two, as it were, skeletal outlines of harmonies, each skeletal outline determined by the two extreme numbers wherewith the limits

[1]That is to say, since Saturn and Jupiter have one revolution with respect to one another every twenty years, they are 81° apart once every twenty years, while the end-positions of this 81° interval traverse the ecliptic in leaps, so to speak, and coincide with the apsides approximately once in eight hundred years. C. G. W.

of the tunings are designated, and let us search out what fits in with them from the variety of movements granted to each planet.

Harmonies of all the Planets, or Universal Harmonies in the Major Mode

In order that b may be in concord			At gravest Tuning	At most acute Tuning	[Modern notation
☿	e⁷		380′20″		5 x 8va
	b⁶		285′15″	292′48″	
	g⁶		228′12″	234′16″	
♀	e⁶		190′10″	195′14″	4 x 8va
	e⁵		95′5″	97′37″	
☽	g⁴		57′3″	58′34″	2 x 8va
	b³		35′39″	36′36″	
♂	g³		28′32″	29′17″	8va
♃	b			4′34″	
♭	B		2′14″		
	G		1′47″	1′49″	

E. C., Jr.]

In order that c may be in concord			At Gravest tuning	At most acute tuning	[Modern notation
☿	e⁷		380′20″		5 x 8va
	c⁷		204′16″	312′21″	
	g⁶		228′12″	234′16″	
♀	e⁶		190′10″	195′14″	4 x 8va
	e⁵		95′5″	97′37″	
☽	g⁴		57′3″	58′34″	2 x 8va
	c⁴		38′2″	39′3″	
♂	g³		28′32″	29′17″	8va
♃	c¹		4′45″	4′53″	
♭	G		1′47″	1′49″	

E. C., Jr.]

Saturn joins in this universal consonance with its aphelial movement, the Earth with its aphelial, Venus approximately with its aphelial; at highest tuning, Venus joins with its perihelial; at mean tuning, Saturn joins with its perihelial, Jupiter with its aphelial, Mercury with its perihelial. So Saturn can join in with two movements, Mars with two, Mercury with four. But with the rest remaining, the perihelial movement of Saturn and the aphelial of Jupiter are not allowed. But in their place, Mars joins in with perihelial movement.

The remaining planets join in with single movements, Mars alone with two, and Mercury with four.

[297] Accordingly, the second skeletal outline will be that wherein the other possible consonance, 5 : 8, exists between the Earth and Venus. Here one eighth of the 94′50″ of the diurnal aphelial movement of Venus or 11′51″+, if multiplied by 5, equals the 59′16″ of the movement of the Earth; and similar parts of the 97′37″ of the perihelial movement of Venus are equal to the 61′1″ of the movement of the Earth. Accordingly, the other planets are in concord in the following diurnal movements:

Harmonies of all the Planets, or Universal Harmonies in the Minor Mode

In order that ♭ may be in concord		*At gravest tuning*	*At most acute tuning*	[*Modern notation:*
♄	e♭⁷	379′20″		5 x 8va
	b♭⁷	284′32″	295′56″	
	g⁶	237′4″	244′4″	
♀	e♭⁶	189′40″	195′14″	4 x 8va
	e♭⁵	94′50″	97′37″	
♁	g⁴	59′16″	61′1″	2 x 8va
	b♭⁴	35′35″	36′37″	
♂	g³	29′38″	30′31″	8va
♃	b♭¹			
			4′35″	
♭	b♭	2′13″		
	G	3′51″	1′55″	

E. C., Jr.]

Here again, in the mean tuning Saturn joins in with its perihelial movement, Jupiter with its aphelial, Mercury with its perihelial. But at highest tuning approximately the perihelial movement of the Earth joins in.

In order that c may be in concord	At gravest tuning	At most acute tuning	[Modern notation:
☿ eb7 / c7 / g6	379'20" / 316'5" / 237'4"	— / 325'26" / 244'4"	5 x 8va
♀ eb6 / c6 / eb5	189'40" / — / 94'50"	195'14" / 162'43" / 97'37"	4 x 8va
♁ g4	59'16"	61'1"	2 x 8va
♂ g3	29'38"	30'31"	8va
♃ c1	4'56"	5'5"	
♄ G	3'51"	1'55"	

E. C., Jr.]

An here, with the aphelial movement of Jupiter and the perihelial movement of Saturn removed, the aphelial movement of Mercury is practically admitted besides the perihelial. The rest remain.

Therefore astronomical experience bears witness that the universal consonances of all the movements can take place, and in the two modes [*generum*], the major and minor, and in both genera of form, or (if I may say so) in respect to two pitches and in any one of the four cases, with a certain latitude of tuning and also with a certain variety in the particular consonances of Saturn, Mars, and Mercury, of each with the rest; and that is not afforded by the intermediate movements alone, but by all the extreme movements too, except the aphelial movement of Mars and the perihelial movement of Jupiter; because since the former occupies *f* sharp; and the latter, *d* Venus, which occupies perpetually the intermediate *e* flat or *e*, does not allow those neighbouring dissonances in the universal consonance, as she would do if she had space to go beyond *e* or *e* flat. This difficulty is caused by the wedding of the Earth and Venus, or the male and the female. These two planets divide the kinds [*genera*] of consonances into the major and masculine and the minor and feminine, according as the one spouse has gratified the other—namely, either the Earth is in its aphelion, as if preserving [298] its marital dignity and performing works worthy of a man, with

Venus removed and pushed away to her perihelion as to her distaff; or else the Earth has kindly allowed her to ascend into aphelion or the Earth itself has descended into its perihelion towards Venus and as it were, into her embrace, for the sake of pleasure, and has laid aside for a while its shield and arms and all the works befitting a man; for at that time the consonance is minor.

But if we command this contradictory Venus to keep quiet, *i.e.*, if we consider what the consonances not of all but merely of the five remaining planets can be, excluding the movement of Venus, the Earth still wanders around its *g* string and does not ascend a semitone above it. Accordingly *b♭*, *b*, *c*, *d*, *e♭*, and *e* can be in concord with *g*, whereupon, as you see, Jupiter, marking the *d* string with its perihelial movement, is brought in. Accordingly, the difficulty about Mars' aphelial movement remains. For the aphelial movement of the Earth, which occupies *g*, does not allow it on *f* sharp; but the perihelial movement, as was said above in Chapter v, is in discord with the aphelial movement of Mars by about half a diesis.

Harmonies of the Five Planets, with Venus Left Out

Major mode (Genus durum)		At gravest tuning	At most acute tuning	[Modern notation:
♄	d⁷	342′18″	351′24″	5 x 8va
	b⁶	285′15″	292′48″	
	g⁶	228′12″	234′16″	
♀ in discord	d⁶	171′9″	175′42″	4 x 8va
	e⁵	95′5″	97′37″	
♂	g⁴	57′3″	58′34″	2 x 8va
	b³	35′39″	36′36″	
	g³	28′31″	29′17″	8va
♃	d¹	5′21″	5′30″	
	b¹		4′35″	
♭ B		2′13″		
G		1′47″		

E. C., Jr.]

Here at the most grave tuning, Saturn and the Earth join in with their aphelial movements; at the mean tuning, Saturn with its perihelial and Jupiter with its aphelial; at the most acute, Jupiter with its perihelial.

Minor mode (Genus molle)			At gravest tuning	At most acute tuning	[Modern notation:
					5 x 8va
☿	d⁷ b⁶ g⁶		342′18″ 273′50″ 228′12″	351′24″ 280′57″ 234′16″	
♀ in discord	d⁶ e⁵		171′9″ 95′5″	175′42′ 97′37″	4 x 8va
♂	g⁴ b³		57′3″ 34′14″	58′34″ 35′8″	2 x 8va
	g³		28′31″	29′17″	8va
♃	d¹		5′21″	5′30″	
♭	B G		2′8″ 1′47″	2′12″ 1′50″	

E. C., Jr.]

Here the aphelial movement of Jupiter is not allowed, but at the most acute tuning Saturn practically joins in with its perihelial movement.

But there can also exist the following harmony of the four planets, Saturn, Jupiter, Mars, and Mercury, wherein too the aphelial movement of Mars is present, but it is without latitude of tuning.

In order that b maybe in concord   [*Modern notation:*

𝄵 d⁷		335′50″
b⁶		279′52″
f#⁶		209′52″
d⁶		167′55″
♂ b³		34′59″
f#³		26′14″
♃ d¹		5′15″
♭ B		2′11″

E. C., Jʀ.]

In order that a may be in concord   [*Modern notation:*

𝄵 d⁷	
a⁶	
f#⁶	
d⁶	
♂ a³	
f#³	
♃ d¹	
♭ A	

E. C., Jʀ.]

Accordingly the movements of the heavens are nothing except a certain ever-lasting polyphony (intelligible, not audible) with dissonant tunings, like certain syncopations or cadences (wherewith men imitate these natural dissonances), which tends towards fixed and prescribed clauses—the single clauses having six terms (like voices)— and which marks out and distinguishes the immensity of time with those notes. Hence it is no longer a surprise that man, the ape of his Creator, should finally have discovered the art of singing polyphonically [*per concentum*], which was unknown to the ancients, namely in order that he might play the everlastingness of all created time in some short part of an hour by means of an artistic concord of many voices and that he might to some extent taste the satisfaction of God the Workman with His own works, in that very sweet sense of delight elicited from this music which imitates God.

NOTE: The comparison Kepler draws between the celestial harmonies and the polyphonic music of his time may be clarified by a simple example for four voices from—Palestrina, *O Crux:*

X *Consonant harmonies*

Y *Dissonant syncopations*

Z *Resolutions of dissonances*

Cadence

As will be observed each of the four voices (as it would also be with the six to which **Kepler** refers) moves from one consonant chord to another while following a graceful melodic line. Sometimes bits of scales or passing tones are added to give a voice more melodic freedom expressiveness. For the same reason a voice may remain on the same note while the other voices change to a new chord. When this becomes a dissonance (called a syncopation) in the new chord it usually resolves by moving one step downward to a tone that is consonant with the other voices. As in this example each section or "caluse" ends with a cadence.

E. C., JR.

8. In the Celestial Harmonies Which Planet Sings Soprano, Which Alto, Which Tenor, and Which Bass?

Although these words are applied to human voices, while voices or sounds do not exist in the heavens, on account of the very great tranquillity of movements, and not even the subjects in which we find the consonances are comprehended under the true genus of movement, since we were considering the movements solely as apparent from the sun, and finally, although there is no such cause in the heavens, as in human singing, for requiring a definite number of voices in order to make consonance (for first there was the number of the six planets revolving around the sun, from the number of the five intervals taken from the regular figures, and then afterwards—in the order of nature, not of time—the congruence of the movements was settled): I do not know why but nevertheless this wonderful congruence with human song has such a strong effect upon me that I am compelled to pursue this part of the comparison, also, even without any solid natural cause. For those same properties which in Book III, [300] Chapter 16, custom ascribed to the bass and nature gave legal grounds for so doing are somehow possessed by Saturn and Jupiter in the heavens; and we find those of the tenor in Mars, those of the alto are present in the Earth and Venus, and those of the soprano are possessed by Mercury, if not with equality of intervals, at least proportionately. For howsoever in the following chapter the eccentricities of each planet are deduced from their proper causes and through those eccentricities the intervals proper to the movements of each, none the less there comes from that the following wonderful result (I do not know whether it is occasioned by the procurement and mere tempering of necessities): (1) as the bass is opposed to the alto, so there are two planets which have the nature of the alto, two that of the bass, just as in any Mode of song there is one [bass and one alto] on either side, while there are single representatives of the other single voices. (2) As the alto is practically supreme in a very narrow range [in angustiis] on account of necessary and natural causes unfolded in Book III, so the almost innermost planets, the Earth and Venus, have the narrowest intervals of movements, the Earth not much more than a semitone, Venus not even a diesis. (3) And as the tenor is free, but none the less progresses with moderation, so Mars alone—with the single exception of Mercury—can make the greatest interval, namely a perfect fifth. (4) And as the bass makes harmonic leaps, so Saturn and Jupiter have intervals which are harmonic, and in relation to one another pass from the octave to the octave and perfect fifth. (5) And as the soprano is the freest, more than all the rest, and likewise the swiftest, so Mercury can traverse more than an octave in the shortest period. But this is altogether *per accidens;* now let us hear the reasons for the eccentricities.

9. The Genesis of the Eccentricities in the Single Planets from the Procurement of the Consonances between their Movements

Accordingly, since we see that the universal harmonies of all six planets cannot take place by chance, especially in the case of the extreme movements, all of which we see concur in the universal harmonies—except two, which concur in harmonies closest to the universal—and since much less can it happen by chance that all the pitches of the system of the octave (as set up in Book III) by means of harmonic divisions are designated by the extreme planetary movements, but least of all that the very subtle business of the distinction of the celestial consonances into two modes, the major and minor, should be the outcome of chance, without the special attention of the Artisan: accordingly it follows that the Creator, the source of all wisdom, the everlasting approver of order, the eternal and superexistent geyser of geometry and harmony, it follows, I say, that He, the Artisan of the celestial movements Himself, should have conjoined to the five regular solids the harmonic ratios arising from the regular plane figures, and out of both classes should have formed one most perfect archetype of the heavens: in order that in this archetype, as through the five regular solids the shapes of the spheres shine through on which the six planets are carried, so too through the consonances, which are generated from the plane figures, and deduced from them in Book III, the measures of the eccentricities in the single planets might be determined so as to proportion the movements of the planetary bodies; and in order that there should be one tempering together of the ratios and the consonances, and that the greater ratios of the spheres should yield somewhat to the lesser ratios of the eccentricities necessary for procuring the consonances, and conversely those in especial of the harmonic ratios which had a greater kinship with each solid figure should be adjusted to the planets— in so far as that could be effected by means of consonances. And in order that, finally, in that way both the ratios of the spheres and the eccentricities of the single planets might be born of the archetype simultaneously, while from the amplitude of the spheres and the bulk of the bodies the periodic times of the single planets might result.

[301] While I struggle to bring forth this process into the light of human intellect by means of the elementary form customary with geometers, may the Author of the heavens be favourable, the Father of intellects, the Bestower of mortal senses, Himself immortal and superblessed, and may He prevent the darkness of our mind from bringing forth in this work anything unworthy of His Majesty, and may He effect that we, the imitators of God by the help of the Holy Ghost, should rival the perfection of His works in sanctity of life, for which He choose His church throughout the Earth and, by the blood of His Son, cleansed it from sins, and that we should keep at a distance all the discords of enmity, all contentions, rivalries, anger, quarrels, dissensions, sects, envy, provocations, and irritations arising through mocking speech and the other works of the flesh; and that along with myself, all who possess the spirit of Christ will not only desire but will also strive by deeds to express and make sure their calling, by spurning all crooked morals of all kinds which have been veiled and painted over with the cloak of zeal or of the love of truth or of singular erudition or modesty over against contentious teachers, or with any other showy garment. Holy Father, keep us safe in the concord of our love for one another, that we may be one, just as Thou art one with They Son, Our Lord, and with the Holy Ghost, and just as through the sweetest bonds of harmonies Thou hast made all Thy works one; and that from the bringing of Thy people into concord the body of Thy Church may be beuilt up in the Earth, as Thou didst erect the heavens themselves out of harmonies.

HARMONIES OF THE WORLD

I. AXIOM. *It is reasonable that, wherever in general it could have been done, all possible harmonies were due to have been set up between the extreme movements of the planets taken singly and by twos, in order that that variety should adorn the world.*

II. AXIOM *The five intervals between the six spheres to some extent were due to correspond to the ratio of the geometrical spheres which inscribe and circumscribe the five regular solids, and in the same order which is natural to the figures.*

Concerning this, see Chapter 1 and the *Mysterium Cosmographicum* and the *Epitome of Copernican Astronomy.*

III. PROPOSITION. *The intervals between the Earth and Mars, and between the Earth and Venus, were due to be least, in proportion to their spheres, and thereby approximately equal; middling and approximately equal between Saturn and Jupiter, and between Venus and Mercury; but greatest between Jupiter and Mars.*

For by Axiom II, the planets corresponding in position to the figures which make the least ratio of geometrical spheres ought likewise to make the least ratio; but those which correspond to the figures of middling ratio ought to make the greatest; and those which correspond to the figures of greatest ratio, the greatest. But the order holding between the figures of the dodechahedron and the icosahedron is the same as that between the pairs of planets, Mars and the Earth, and the Earth and Venus, and the order of the cube and octahedron is the same as that of the pair Saturn and Jupiter and that of the pair Venus and Mercury; and, finally, the order of the tetrahedron is the same as that of the pair Jupiter and Mars (see Chapter 3). Therefore, the least ratio will hold between the planetary spheres first mentioned, while that between Saturn and Jupiter is approximately equal to that between Venus and Mercury; and, finally, the greatest between the spheres of Jupiter and Mars.

IV. AXIOM. *All the planets ought to have their eccentricities diverse, no less than a movement in latitude, and in proportion to those eccentricities also their distances from the sun, the source of movement, diverse.*

As the essence of movement consists not in *being* but in *becoming*, so too the form or figure of the region which any planet traverses in its movement does not become solid immediately from the start but in the succession of time acquires at last not only length but also breadth and depth (its perfect ternary of dimensions); and, gradually, thus, by the interweaving and piling up of many circuits, the form of a concave sphere comes to be represented—just as out of the silk-worm's thread, by the interweaving and heaping together of many circles, the cocoon is built.

V. PROPOSITION. *Two diverse consonances were to have been attributed to each pair of neighbouring planets.*

For, by Axiom IV, any planet has a longest and a shortest distance from the sun, wherefore, by Chapter 3, it will have both a slowest movement and a fastest. Therefore, there are two primary comparisons of the extreme movements, one of the diverging movements in the two planets, and the other of the converging. Now it is necessary that they be diverse from one another, because the ratio of the diverging movements will be greater, that of the converging, lesser. But, moreover, diverse consonances had to exist by way of diverse pairs of planets, so that this variety should make for the adornment of the world—by Axiom I—and also because the ratios of the intervals between two planets are diverse, by Proposition III. But to each definite ratio of the spheres there correspond harmonic ratios, in quantitative kinship, as has been demonstrated in Chapter 5 of this book.

VI. PROPOSITION. *The two least consonances, 4 : 5 and 5 : 6, do not have a place between two planets.*

For

$$5 : 4 = 1,000 : 800$$

and

$$6 : 5 = 1,000 : 833.$$

But the spheres circumscribed around the dodecahedron and icosahedron have a greater ratio to the inscribed spheres than 1,000 : 795, etc., and these two ratios indicate the intervals between the nearest planetary spheres, or the least distances. For in the other regular solids the spheres are farther distant from one another. But now the ratio of the movements is even greater than the ratios of the intervals, unless the ratio of the eccentricities to the spheres is vast —by Article XIII of Chapter 3. Therefore the least ratio of the movements is greater than 4 : 5 and 5 : 6. Accordingly, these consonances, being hindered by the regular solids, receive no place among the planets.

VII. PROPOSITION. *The consonance of the perfect fourth can have no place between the converging movements of two planets, unless the ratios of the extreme movements proper to them are, if compounded, more than a perfect fifth.*

For let 3 : 4 be the ratio between the converging movements. And first, let there be no eccentricity, no ratio of movements proper to the single planets, but both the converging and the mean movements the same; then it follows that the corresponding intervals, which by this hypothesis will be the semidiameters of the spheres, constitute the ⅔d power of this ratio, *viz.*, 4480 : 5424 (by Chapter 3). But this ratio is already less than the ratio of the spheres of any regular figure; and so the whole inner sphere would be cut by the regular planes of the figure inscribed in any outer sphere. But this is contrary to Axiom II.

Secondly, let there be some composition of the ratios between the extreme movements, and let the ratio of the converging movements be 3 : 4 or 75 : 100, but let the ratio of the corresponding intervals be 1,000 : 795, since no regular figure has a lesser ratio of spheres. And because the inverse ratio of the movements exceeds this ratio of the intervals by the excess 750 : 795, then if this excess is divided into the ratio 1,000 : 795, according to the doctrine of Chapter 3, the result will be 9434 : 7950, the square root of the ratio of the spheres. Therefore the square of this ratio, *viz.*, 8901 : 6320, *i.e.*, 10,000 : 7,100 is the ratio of the spheres. Divide this by 1000 : 795, the ratio of the converging intervals, the result will be 7100 : 7950, about a major whole tone. The compound of the two ratios which the mean movements have to the converging movements on either side must be at least so great, in order that the perfect fourth may be possible between the converging movements. Accordingly, the compound ratio of the diverging extreme intervals to the converging extreme intervals is about the square root of this ratio, *i.e.*, two tones, and again the converging intervals are the square of this, *i.e.*, more than a perfect fifth. Accordingly, if the compound of the proper movements of two neighbouring planets is less than a perfect fifth, a perfect fourth will not be possible between their converging movements.

VIII. PROPOSITION. *The consonances 1 : 2 and 1 : 3, i.e., the octave and the octave plus a fifth were due to Saturn and Jupiter.*

For they are the first and highest of the planets and have obtained the first figure, the cube, by Chapter 1 of this book; and these consonances are first in the order of nature and are chief in the two families of figures, the bisectorial or tetragonal and the triangular, by what has been said in Book I. But that which is chief, the octave 1 : 2, is approximately greater than the ratio of the spheres of the cube, [303] which is 1 : $\sqrt{3}$; wherefore it is fitted to become the lesser ratio of the movements of the planets on the cube, by Chapter 3, Article XIII; and, as a consequence, 1 : 3 serves as the greater ratio.

But this is also the same as what follows: for if some consonance is to some ratio of the spheres of the figures, as the ratio of the movements apparent from the sun is to the ratio of the mean intervals, such a consonance will duly be attributed to the movements. But it is natural that the ratio of the diverging movements should be much greater than the ratio of the ⅗th powers of the spheres, according to the end of Chapter 3, *i.e.*, it approaches the square of the ratio of the spheres; and moreover 1 : 3 is the square of the ratio of the spheres of the cube, which we call the ratio of 1 : $\sqrt{3}$. Therefore, the ratio of the diverging movements of Saturn and Jupiter is 1 : 3. (See above, Chapter 2, for many other kinships of these ratios with the cube.)

IX. PROPOSITION. *The private ratios of the extreme movements of Saturn and Jupiter compounded were due to be approximately 2 : 3, a perfect fifth.*

This follows from the preceding; if the perihelial movement of Jupiter is triple the aphelial movement of Saturn, and conversely the aphelial movement of Jupiter is double the perihelial of Saturn, then 1 : 2 and 1 : 3 compounded inversely give 2 : 3.

X. AXIOM. *When choice is free in other respects, the private ratio of movements, which is prior in nature or of a more excellent mode or even which is greater, is due to the higher planet.*

XI. PROPOSITION. *The ratio of the aphelial movement of Saturn to the perihelial was due to be 4 : 5, a major third, but that of Jupiter's movements 5 : 6, a minor third.*

For as compounded together they are equivalent to 2 : 3; but 2 : 3 can be divided harmonically no other way than into 4 : 5 and 5 : 6. Accordingly God the composer of harmonies divided harmonically the consonance 2 : 3, (by Axiom I) and the harmonic part of it which is greater and of the more excellent major mode, as masculine, He gave to Saturn the greater and higher planet, and the lesser ratio 5 : 6 to the lower one, Jupiter (by Axiom X).

XII. PROPOSITION. *The great consonance of 1 : 4, the double octave, was due to Venus and Mercury.*

For as the cube is the first of the primary figures, so the octahedron is the first of the secondary figures, by Chapter 1 of this book. And as the cube considered geometrically is outer and the octahedron is inner, *i.e.*, the latter can be inscribed in the former, so also in the world Saturn and Jupiter are the beginning of the upper and outer planets, or from the outside; and Mercury and Venus are the beginning of the inner planets, or from the inside, and the octahedron has been placed between their circuits: (see Chapter 3). Therefore, from among the consonances, one which is primary and cognate to the octahedron is due to Venus and Mercury. Furthermore, from among the consonances, after 1 : 2 and

1 : 3, there follows in natural order 1 : 4; and that is cognate to 1 : 2, the consonance of the cube, because it has arisen from the same cut of figures, *viz.*, the tetragonal, and is commensurable with it, *viz.*, the double of it; while the octahedron is also akin to, and commensurable with the cube. Moreover, 1 : 4 is cognate to the octahedron for a special reason, on account of the number four being in that ratio, while a quadrangular figure lies concealed in the octahedron and the ratio of its spheres is said to be 1 : $\sqrt{2}$.

Accordingly the consonance 1 : 4 is a continued power of this ratio, in the ratio of the squares, *i.e.*, the 4th power of 1 : $\sqrt{2}$ (see Chapter 2). Therefore, 1 : 4 was due to Venus and Mercury. And because in the cube 1 : 2 has been made the smaller consonance of the two, since the outermost position is over against it, in the octahedron there will be 1 : 4, the greater consonance of the two, as the innermost position is over against it. But too, this is the reason why 1 : 4 has here been given as the greater consonance, not as the smaller.[1] For since the ratio of the spheres of the octahedron is the ratio of 1 : $\sqrt{3}$, then if it is postulated that the inscription of the octahedron among the planets is perfect (although it is not perfect, but penetrates Mercury's sphere to some extent— which is of advantage to us): accordingly, the ratio of the converging movements must be less than the $\frac{3}{2}$th powers of 1 : $\sqrt{3}$; but indeed 1 : 3 is plainly the square of the ratio 1 : $\sqrt{3}$ and is thus greater than the exact ratio; all the more then will 1 : 4 be greater than the exact ratio, as greater than 1 : 3. Therefore, not even the square root of 1 : 4 is allowed between the converging movements. Accordingly, 1 : 4 cannot be less than the octahedric; so it will be greater.

Further: 1 : 4 is akin to the octahedric square, where the ratio of the inscribed and circumscribed circles is 1 : $\sqrt{2}$, just as 1 : 3 is akin to the cube, where the ratio of the spheres is 1 : $\sqrt{3}$. For as 1 : 3 is a power of 1 : $\sqrt{3}$, *viz.*, its square, [304] so too here 1 : 4 is a power of 1 : $\sqrt{2}$, *viz.*, twice its square, *i.e.*, its quadruple power. Wherefore, if 1 : 3 was due to have been the greater consonance of the cube (by Proposition VII), accordingly 1 : 4 ought to become the greater consonance of its octahedron.

XIII. PROPOSITION. *The greater consonance of approximately 1 : 8, the triple octave, and the smaller consonance of 5 : 24, the minor third and double octave, were due to the extreme movements of Jupiter and Mars.*

For the cube has obtained 1 : 2 and 1 : 3, while the ratio of the spheres of the tetrahedron, which is situated between Jupiter and Mars, called the triple ratio, is the square of the ratio of the spheres of the cube, which is called the ratio of 1 : $\sqrt{3}$. Therefore, it was proper that ratios of movements which are the squares of the cubic ratios should be applied to the tetrahedron. But of the ratios 1 : 2 and 1 : 3 the following ratios are the squares: 1 : 4 and 1 : 9. But 1 : 9 is not harmonic, and 1 : 4 has already been used up in the octahedron. Accordingly, consonances neighbouring upon these ratios were to have been taken, by Axiom I. But the lesser ratio 1 : 8 and the greater 1 : 10 are the nearest. Choice between these ratios is determined by kinship with the tetrahedron, which has nothing in common with the pentagon, since 1 : 10 is of a pentagonal cut, but the tetrahedron has greater kinship with 1 : 8 for many reasons (see Chapter 2).

Further, the following also makes for 1 : 8: just as 1 : 3 is the greater consonance of the cube and 1 : 4 the greater consonance of the octahedron, because

[1] *Smaller* (lesser) and *greater* consonances are equivalent to our modern "more closely spaced" and "more widely spaced" consonances. E. C., Jr.

they are powers of the ratios between the spheres of the figures, so too 1 : 8 was due to be the greater consonance of the tetrahedron, because as its body is double that of the octahedron inscribed in it, as has been said in Chapter 1, so too the term 8 in the tetrahedral ratio is double the term 4 in the tetrahedral ratio.

Further, just as 1 : 2 the smaller consonance of the cube, is one octave, and 1 : 4, the greater consonance of the octahedron, is two octaves, so already 1 : 8, the greater consonance of the tetrahedron, was due to be three octaves. Moreover, more octaves were due to the tetrahedron than to the cube and octahedron, because, since the smaller tetrahedral consonance is necessarily greater than all the lesser consonances in the other figures (for the ratio of the tetrahedral spheres is greater than all the spheres of figures): too the greater tetrahedral consonance was due to exceed the greater consonances of the others in number of octaves. Finally, the triple of octave intervals has kinship with the triangular form of the tetrahedron, and has a certain perfection, as follows: every three is perfect; since even the octuple, the term [of the triple octave], is the first cubic number of perfect quantity, namely of three dimensions.

A greater consonance neighbouring upon 1 : 4 or 6 : 24 is 5 : 24, while a lesser is 6 : 20 or 3 : 10. But again 3 : 10 is of the pentagonal cut, which has nothing in common with the tetrahedron. But on account of the numbers 3 and 4 (from which the numbers 12, 24 arise) 5 : 24 has kinship with the tetrahedron. For we are here neglecting the other lesser terms, *viz.*, 5 and 3, because their lightest degree of kinship is with figures, as it is possible to see in Chapter 2. Moreover, the ratio of the spheres of the tetrahedron is triple; but the ratio of the converging intervals too ought to be approximately so great, by Axiom II. By Chapter 3, the ratio of the converging movements approaches the inverse ratio of the ⅗th powers of the intervals, but the ⅗th power of 3 : 1 is approximately 1000 : 193. Accordingly, whereof the aphelial movement of Mars is 1000, the [perihelial] of Jupiter will be slightly greater than 193 but much less than 333, which is one third of 1,000. Accordingly, not the consonance 10 : 3, *i.e.*, 1,000 : 333, but the consonance 24 : 5, *i.e.*, 1,000 : 208, takes place between the converging movements of Jupiter and Mars.

XIV. PROPOSITION. *The private ratio of the extreme movements of Mars was due to be greater than 3 : 4, the perfect fourth, and approximately 18 : 25.*

For let there be the exact consonances 5 : 24 and 1 : 8 or 3 : 24, which are commonly attributed to Jupiter and Mars (Proposition XIII). Compound inversely 5 : 24, the lesser with 3 : 24, the greater; 3 : 5 results as the compound of both ratios. But the proper ratio of Jupiter alone has been found to be 5 : 6, in Proposition XI, above. Then compound this inversely with the composition 3 : 5, *i.e.*, compound 30 : 25 and 18 : 30; there results as the proper ratio of Mars 18 : 25, which is greater than 18 : 24 or 3 : 4. But it will become still greater, if, on account of the ensuing reasons, the common greater consonance 1 : 8 is increased.

XV. PROPOSITION. *The consonances 2 : 3, the fifth; 5 : 8, the minor sixth; and 3 : 5, the major sixth were to have been distributed among the converging movements of Mars and the Earth, the Earth and Venus, Venus and Mercury, and in that order.*

For the dodecahedron and the icosahedron, the figures interspaced between Mars, the Earth, and Venus have the least ratio between their circumscribed and inscribed spheres. [305] Therefore from among possible consonances the least are due to them, as being cognate for this reason, and in order that Axiom II may have place. But the least consonances of all, *viz.*, 5 : 6 and 4 : 5, are not possible, by Proposition IV. Therefore, the nearest consonances greater than they, *viz.*, 3 : 4 or 2 : 3 or 5 : 8 or 3 : 5 are due to the said figures.

Again, the figure placed between Venus and Mercury, *viz.*, the octahedron, has the same ratio of its spheres as the cube. But by Proposition VII, the cube received the ocatve as the lesser consonance existing between the converging movements. Therefore, by proportionality, so great a consonance, *viz.*, 1 : 2, would be due to the octahedron as the lesser consonance, if no diversity intervened. But the following diversity intervenes: if compounded together, the private ratios of the single movements of the cubic planets, *viz.*, Saturn and Jupiter, did not amount to more than 2 : 3; while, if compounded, the ratios of the single movements of the octahedral planets, *viz.*, Venus and Mercury will amount to more than 2 : 3, as is apparent easily, as follows: For, as the proportion between the cube and octahedron would require if it were alone, let the lesser octahedral ratio be greater than the ratios here given, and thereby clearly as great as was the cubic ratio, *viz.*, 1 : 2; but the greater consonance was 1 : 4, by Proposition XII. Therefore if the lesser consonance 1 : 2 is divided into the one we have just laid down, 1 : 2, still remains as the compound of the proper movements of Venus and Mercury; but 1 : 2 is greater than 2 : 3 the compound of the proper movements of Saturn and Jupiter; and indeed a greater eccentricity follows upon this greater compound, by Chapter 3, but a lesser ratio of the converging movements follows upon the greater eccentricity, by the same Chapter 3. Wherefore by the addition of a greater eccentricity to the proportion between the cube and the octahedron it comes about that a lesser ratio than 1 : 2 is also required between the converging movements of Venus and Mercury. Moreover, it was in keeping with Axiom I that, with the consonance of the octave given to the planets of the cube, another consonance which is very near (and by the earlier demonstration less than 1 : 2) should be joined to the planets of the octahedron. But 3 : 5 is proximately less than 1 : 2, and as the greatest of the three it was due to the figure having the greatest ratio of its spheres, *viz.*, the octahedron. Accordingly, the lesser ratios, 5 : 8 and 2 : 3 or 3 : 4, were left for the icosahedron and dodecahedron, the figures having a lesser ratio of their spheres.

But these remaining ratios have been distributed between the two remaining planets, as follows. For as, from among the figures, though of equal ratios between their spheres, the cube has received the consonance 1 : 2, while the octahedron the lesser consonance 3 : 5, in that the compound ratio of the private movements of Venus and Mercury exceeded the compound ratio of the private movements of Saturn and Jupiter; so also although the dodecahedron has the same ratio of its spheres as the icosahedron, a lesser ratio was due to it than to the icosahedron, but very close on account of a similar reason, *viz.*, because this figure is between the Earth and Mars, which had a great eccentricity in the foregoing. But Venus and Mercury, as we shall hear in the following, have the least eccentricities. But since the octahedron has 3 : 5, the icosahedron, whose species are in a lesser ratio, has the next slightly lesser, *viz.*, 5 : 8; accordingly, either 2 : 3, which remains, or 3 : 4 was left for the dodecahedron, but more likely 2 : 3, as being nearer to the icosahedral 5 : 8; since they are similar figures.

But 3 : 4 indeed was not possible. For although, in the foregoing, the private ratio of the extreme movements of Mars was great enough, yet the Earth—as has already been said and will be made clear in what follows—contributed its own ratio, which was too small for the compound ratio of both to exceed the perfect fifth. Accordingly, Proposition VII, 3 : 4 could not have place. And all the more so, because—as will follow in Proposition XVII—the ratio of the converging intervals was due to be greater than 1,000 : 795.

XVI. PROPOSITION. *The private ratios of movements of Venus and Mercury, if compounded together, were due to make approximately 5 : 12.*

For divide the lesser harmonic ratio attributed in Proposition XV to this pair jointly into the greater of them, 1 : 4 or 3 : 12, by Proposition XII; there results 5 : 12, the compound ratio of the private movements of both. And so the private ratio of the extreme movements of Mercury alone is less than 5 : 12, the magnitude of the private movement of Venus. Understand this of these first reasons. For below, by the second reasons, through the addition of some variation to the joint consonances of both, it results that only the private ratio of Mercury is perfectly 5 : 12.

XVII. PROPOSITION. *The consonance between the diverging movements of Venus and the Earth could not be less than 5 : 12.*

For in the private ratio of its movements Mars alone has received more than the perfect fourth and more than 18 : 25, by Proposition XIV. But their lesser consonance is the perfect fifth, [306] by Proposition XV. Accordingly, the ratio compounded of these two parts is 12 : 25. But its own private ratio is due to the Earth, by Axiom IV. Therefore, since the consonance of the diverging movements is made up out of the said three elements, it will be greater than 12 : 25. But the nearest consonance greater than 12 : 25, *i.e.*, 60 : 125, is 5 : 12, *viz.*, 60 : 144. Wherefore, if there is need of a consonance for this greater ratio of the two planets, by Axiom I, it cannot be less than 60 : 144 or 5 : 12.

Therefore up to now all the remaining pairs of planets have received their two consonances by necessary reasons; the pair of the Earth and Venus alone has as yet been allotted only one consonance, 5 : 8, by the axioms so far employed. Therefore, we must now take a new start and inquire into its remaining consonance, *viz.*, the greater, or the consonance of the diverging movements.

POSTERIOR REASONS

XVIII. AXIOM. *The universal consonances of movements were to be constituted by a tempering of the six movements, especially in the case of the extreme movements.*

This is proved by Axiom I.

XIX. AXIOM. *The universal consonances had to come out the same within a certain latitude of movements, namely, in order that they should occur the more frequently.*

For if they had been limited to indivisible points of the movements, it could have happened that they would never occur, or very rarely.

XX. AXIOM. *As the most natural division of the kinds [generum] of consonances is into major and minor, as has been proved in Book 3, so the universal consonances of both kinds had to be procured between the extreme movements of the planets.*

XXI. AXIOM. *Diverse species of both kinds of consonances had to be instituted, so that the beauty of the world might well be composed out of all possible forms of*

*variety—and by means of the extreme movements, at least by means of some ex-
treme movements.*

By Axiom I.

XXII. PROPOSITION. *The extreme movements of the planets had to designate
pitches or strings [chordas] of the octave system, or notes [claves] of the musical
scale.*

For the genesis and comparison of consonances beginning from one common
term has generated the musical scale, or the division of the octave into its
pitches or tones [sonos], as has been proved in Book 3. Accordingly, since
varied consonances between the extremes of movements are required, by
Axioms I, XX, and XXI, wherefore the real division of some celestial system
or harmonic scale by the extremes of movements is required.

XXIII. PROPOSITION. *It was necessary for there to be one pair of planets, between the movements of which no consonances could exist except the major sixth
3 : 5 and the minor sixth 5 : 8.*

For since the division into kinds of consonances was necessary, by Axiom XX,
and by means of the extreme movements at the apsides, by XXII, because
solely the extremes, *viz.,* the slowest and the fastest, need the determination
of a manager and orderer, the intermediate tensions come of themselves,
without any special care, with the passage of the planet from the slowest movement to the fastest: accordingly, this ordering could not take place otherwise
than by having the diesis or 24 : 25 designated by the extremes of the two
planetary movements, in that the kinds of consonances are distinguished by
the diesis, as was unfolded in Book 3.

But the diesis is the difference either between two thirds, 4 : 5 and 5 : 6,
or between two sixths, 3 : 5 and 5 : 8, or between those ratios increased by
one or more octave intervals. But the two thirds, 4 : 5 and 5 : 6, did not have
place between two planets, by Proposition VI, and neither the thirds nor the
sixths increased by the interval of an octave have been found, except 5 : 12
in the pair of Mars and the Earth, and still not otherwise than along with the
related 2 : 3, and so the intermediate ratios 5 : 8 and 3 : 5 and 1 : 2 were alike
admitted. Therefore, it remains that the two sixths, 3 : 5 and 5 : 8, were to
be given to one pair of planets. But too the sixths alone were to be granted
to the variation of their movements, in such fashion that they would neither
expand their terms to the proximately greater interval of one octave, 1 : 2,
[307] nor contract them to the narrows of the proximately lesser interval of
the fifth, 2 : 3. For, although it is true that the same two planets, which make
a perfect fifth with their extreme converging movements, can also make
sixths and thus traverse the diesis too, still this would not smell of the singular
providence of the Orderer of movements. For the diesis, the least interval—
which is potentially latent in all the major intervals comprehended by the
extreme movements—is itself at that time traversed by the intermediate
movements varied by continuous tension, but it is not determined by their
extremes, since the part is always less than the whole, *viz.,* the diesis than
the greater interval 3 : 4 which exists between 2 : 3 and 1 : 2 and which whole
would be here assumed to be determined by the extreme movements.

XXIV. PROPOSITION. *The two planets which shift the kind [genus] of harmony,
which is the difference between the private ratios of the extreme movements, ought
to make a diesis, and the private ratio of one ought to be greater than a diesis,*

and they ought to make one of the sixths with their aphelial movements and the other with their perihelial.

For, since the extremes of the movements make two consonances differing by a single diesis, that can take place in three ways. For either the movement of one planet will remain constant and the movement of the other will vary by a diesis, or both will vary by half a diesis and make 3 : 5, a major sixth, when the upper is at its aphelion and the lower in its perihelion, and when they move out of those intervals and advance towards one another, the upper into its perihelion and the lower into its aphelion, they make 5 : 8, a minor sixth; or, finally, one varies its movement from aphelion to perihelion more than the other does, and there is an excess of one diesis, and thus there is a major sixth between the two aphelia, and a minor sixth between the two perihelia. But the first way is not legitimate, for one of these planets would be without eccentricity, contrary to Axiom IV. The second way was less beautiful and less expedient; less beautiful, because less harmonic, for the private ratios of the movements of the two planets would have been out of tune [*inconcinnae*], for whatever is less than a diesis is out of tune; moreover it occasions one single planet to labour under this ill-concordant small difference—except that indeed it could not take place, because in this way the extreme movements would have wandered from the pitches of the system or the notes [*clavibus*] of the musical scale, contrary to Proposition XXII. Moreover, it would have been less expedient, because the sixths would have occurred only at those moments in which the planets would have been at the contrary apsides; there would have been no latitude within which these sixths and the universal consonances related to them could have occurred; accordingly, these universal consonances would have been very rare, with all the [*harmonic*] positions of the planets reduced to the narrow limits of definite and single points on their orbits, contrary to Axiom XIX. Accordingly, the third way remains: that both of the planets should vary their own private movements, but one more than the other, by one full diesis at the least.

XXV. PROPOSITION. *The higher of the planets which shift the kind of harmony ought to have the ratio of its private movements less than a minor whole tone 9 : 10; while the lower, less than a semitone 15 : 16.*

For they will make 3 : 5 either with their aphelial movements or with their perihelial, by the foregoing proposition. Not with their perihelial, for then the ratio of their aphelial movements would be 5 : 8. Accordingly, the lower planet would have its private ratio one diesis more than the upper would, by the same foregoing proposition. But that is contrary to Axiom X. Accordingly, they make 3 : 5 with their aphelial movements, and with their perihelial 5 : 8, which is 24 : 25 less than the other. But if the aphelial movements make 3 : 5, a major sixth, therefore, the aphelial movement of the upper together with the perihelial of the lower will make more than a major sixth; for the lower planet will compound directly its full private ratio.

In the same way, if the perihelial movements make 5 : 8, a minor sixth, the perihelial movement of the upper and the aphelial movement of the lower will make less than a minor sixth; for the lower planet will compound inversely its full private ratio. But if the private ratio of the lower equalled the semitone 15 : 16, then too a perfect fifth could occur over and above the sixths, because the minor sixth, diminished by a semitone, because the perfect fifth; but this is contrary to Proposition XXIII. Accordingly, the lower planet has less than a semitone in its own interval. And because the private ratio of the upper is one diesis greater than the private ratio of the lower, but the diesis compounded with the semitone makes 9 : 10 the minor whole tone.

XXVI. PROPOSITION. *On the planets which shift the kind of harmony, the upper was due to have either a diesis squared, 576 : 625, i.e., approximately 12 : 13, as* [308] *the interval made by its extreme movements, or the semitone 15 : 16, or something intermediate differing by the comma 80 : 81 either from the former or the latter; while the lower planet, either the simple diesis 24 : 25, or the difference between a semitone and a diesis, which is 125 : 128, i.e., approximately 42 : 43; or, finally and similarly, something intermediate differing either from the former or from the latter by the comma 80 : 81, viz., the upper planet ought to make the diesis squared diminished by a comma, and the lower, the simple diesis diminished by a comma.*

For, by Proposition XXV, the private ratio of the upper ought to be greater than a diesis, but by the preceding proposition less than the [minor] whole tone 9 : 10. But indeed the upper planet ought to exceed the lower by one diesis, by Proposition XXIV. And harmonic beauty persuades us that, even if the private ratios of these planets cannot be harmonic, on account of their smallness, they should at least be from among the concordant [*ex concinnis*] if that is possible, by Axiom I. But there are only two concords less than 9 : 10, the [minor] whole tone, *viz.*, the semitone and the diesis; but they differ from one another not by the diesis but by some smaller interval, 125 : 128. Accordingly, the upper cannot have the semitone; nor the lower, the diesis; but either the upper will have the semitone 15 : 16, and the lower, 125 : 128, *i.e.*, 42 : 43; or else the lower will have the diesis 24 : 25, but the upper the diesis squared, approximately 12 : 13. But since the laws of both planets are equal, therefore, if the nature of the concordant had to be violated in their private ratios, it had to be violated equally in both, so that the difference between their private intervals could remain an exact diesis, which is necessary for distinguishing the kinds of consonances, by Proposition XXIV. But the nature of the concordant was then violated equally in both, if the interval whereby the private ratio of the upper planet fell short of the diesis squared and exceeded the semitone is the same interval whereby the private ratio of the lower planet fell short of a simple diesis and exceeded the interval 125 : 128.

Furthermore, this excess or defect was due to be the comma 80 : 81, because, once more, no other interval was designated by the harmonic ratios, and in order that the comma might be expressed among the celestial movements as it is expressed in harmonics, namely, by the mere excess and defect of the intervals in respect to one another. For in harmonics the comma distinguishes between major and minor whole tones and does not appear in any other way.

It remains for us to inquire which ones of the intervals set forth are preferable —whether the diesis, the simple diesis for the lower planet and the diesis squared for the upper, or the semitone for the upper and 125 : 128 for the lower. And the dieses win by the following arguments: For although the semitone has been variously expressed in the musical scale, yet its allied ratio 125 : 128 has not been expressed. On the other hand, the diesis has been expressed variously and the diesis squared somehow, *viz.*, in the resolution of whole tones into dieses, semitones, and lemmas; for then, as has been said in Book III, Chapter 8, two dieses proximately succeed one another in two pitches. The other argument is that in the distinction into kinds, the laws of the diesis are proper but not at all those of the semitone. Accordingly, there had to be greater consideration of the diesis than of the semitone. It is inferred from everything that the private ratio of the upper planet ought to be 2916 : 3125 or approximately 14 : 15, and that of the lower, 243 : 250 or approximately 35 : 36.

It is asked whether the Highest Creative Wisdom has been occupied in making these tenuous little reckonings. I answer that it is possible that many reasons are hidden from me, but if the nature of harmony has not allowed weightier reasons—since we are dealing with ratios which descend below the magnitude of all concords—it is not absurd that God has followed even those reasons, wherever they appear tenuous, since He has ordained nothing without cause. It would be far more absurd to assert that God has taken at random these magnitudes below the limits prescribed for them, the minor whole tone; and it is not sufficient to say: He took them of that magnitude because He chose to do so. For in geometrical things, which are subject to free choice, God chose nothing without a geometrical cause of some sort, as is apparent in the edges of leaves, in the scales of fishes, in the skins of beasts and their spots and the order of the spots, and similar things.

XXVII. PROPOSITION. *The ratio of movements of the Earth and Venus ought to have been greater than a major sixth between the aphelial movements; less than a minor sixth between the perihelial movements.*

By Axiom XX it was necessary to distinguish the kinds of consonances. But by Proposition XXIII that could not be done except through the sixths. Accordingly, since by Proposition XV the Earth and Venus, planets next to one another and icosahedral, had received the minor sixth, 5 : 8, it was necessary for the other sixth, 3 : 5, to be assigned to them, but not between the converging or diverging extremes, but between the extremes of the same field, one sixth [309] between the aphelial, and the other between the perihelial, by Proposition XXIV. Furthermore, the consonance 3 : 5 is cognate to the icosahedron, since both are of the pentagonal cut. See Chapter 2.

Behold the reason why exact consonances are found between the aphelial and perihelial movements of these two planets, but not between the converging, as in the case of the upper planets.

XXVIII. PROPOSITION. *The private ratio of movements fitting the Earth was approximately 14 : 15, Venus, approximately 35 :36.*

For these two planets had to distinguish the kinds of consonances, by the preceding proposition; therefore, by Proposition XXVI, the Earth as the higher was due to receive the interval 2916 : 3125, *i.e.*, approximately 14 : 15, but Venus as the lower the interval 243 : 250, *i.e.*, approximately 35 : 36.

Behold the reason why these two planets have such small eccentricities and, in proportion to them, small intervals or private ratios of the extreme movements, although nevertheless the next higher planet, Mars, and the next lower, Mercury, have marked eccentricities and the greatest of all. And astronomy confirms the truth of this; for in Chapter 4 the Earth clearly had 14 : 15, but Venus 34 : 35, which astronomical certitude can barely discern from 35 : 36 in this planet.

XXIX. PROPOSITION. *The greater consonance of the movements of Mars and the*

222 KEPLER

Earth, viz., that of the diverging movements, could not be from among the consonances greater than 5 : 12.

Above, in Proposition XVII, it was not any one of the lesser ratios; but now it is not any one of the greater ratios either. For the other common or lesser consonance of these two planets is 2 : 3, when the private ratio of Mars, which by Proposition XIV exceeds 18 : 25, makes more than 12 : 25, *i.e.*, 60 : 125. Accordingly, compound the private ratio of the Earth 14 : 15, *i.e.*, 56 : 60, by the preceding proposition. The compound ratio is greater than 56 : 125, which is approximately 4 : 9, *viz.*, slightly greater than an octave and a major whole tone. But the next greater consonance than the octave and whole tone is 5 : 12, the octave and minor third.

Note that I do not say that this ratio is neither greater nor smaller than 5 : 12; but I say that if it is necessary for it to be harmonic, no other consonance will belong to it.

XXX. PROPOSITION. *The private ratio of movements of Mercury was due to be greater than all the other private ratios.*

For by Proposition XVI the private movements of Venus and Mercury compounded together were due to make about 5 : 12. But the private ratio of Venus, taken separately, is only 243 : 250, *i.e.*, 1458 : 1500. But if it is compounded inversely with 5 : 12, *i.e.*, 625 : 1500, Mercury singly is left with 625 : 1458, which is greater than an octave and a major whole tone; although the private ratio of Mars, which is the greatest of all those among the remaining planets, is less than 2 : 3, *i.e.*, the perfect fifth.

And thereby the private ratios of Venus and Mercury, the lowest planets, if compounded together, are approximately equal to the compounded private ratios of the four higher planets, because, as will now be apparent immediately, the compounded private ratios of Saturn and Jupiter exceed 2 : 3; those of Mars fall somewhat short of 2 : 3: all compounded, 4 : 9, *i.e.*, 60 : 135. Compound the Earth's 14 : 15, *i.e.*, 56 : 60, the result will be 56 : 135, which is slightly greater than 5 : 12, which just now was the compound of the private ratios of Venus and Mercury. But this has not been sought for nor taken from any separate and singular archetype of beauty but comes of itself, by the necessity of the causes bound together by the consonances hitherto established.

XXXI. PROPOSITION. *The aphelial movement of the Earth had to harmonize with the aphelial movement of Saturn, through some certain number of octaves.*

For, by Proposition XVIII, it was necessary for there to be universal consonances, wherefore also there had to be a consonance of Saturn with the Earth and Venus. But if one of the extreme movements of Saturn had harmonized with neither of the Earth's and Venus', this would have been less harmonic than if both of its extreme movements had harmonized with these planets, by Axiom I. Therefore both of Saturn's extreme movements had to harmonize, the aphelial with one of these two planets, the perihelial with the other, since nothing would hinder, as was the case with the first planet. Accordingly these consonances will be either identisonant[1] [*identisonae*] or diversisonant [*diversisonae*], *i.e.*, either of continued double proportion or of some other. But both of them cannot be of some other proportion, for between the terms 3 : 5 (which determine the greater consonance between the aphelial movements of the Earth and Venus, by Proposition XXVII) two harmonic means cannot be set up; for the sixth cannot be

[1] "Identisonant consonances" are such as 3 : 5, 3 : 10, 3 : 20, etc.

divided into three intervals (see Book III). Accordingly, Saturn could not, [310] by means of both its movements, make an octave with the harmonic means between 3 and 5; but in order that its movements should harmonize with the 3 of the earth and the 5 of Venus, it is necessary that one of those terms should harmonize identically, or through a certain number of octaves, with the others, viz., with one of the said planets. But since the identisonant consonances are more excellent, they had to be established between the more excellent extreme movements, viz., between the aphelial, because too they have the position of a principle on account of the altitude of the planets and because the Earth and Venus claim as their private ratio somehow and as a prerogative the consonance 3 : 5, with which as their greater consonance we are now dealing. For although, by Proposition XXII, this consonance belongs to the perihelial movement of Venus and some intermediate movement of the Earth, yet the start is made at the extreme movements and the intermediate movements come after the beginnings.

Now, since on one side we have the aphelial movement of Saturn at its greatest altitude, on the other side the aphelial movement of the Earth rather than Venus is to be joined with it, because of these two planets which distinguish the kinds of harmony, the Earth, again, has the greater altitude. There is also another nearer cause: the posterior reasons—with which we are now dealing—take away from the prior reasons but ony with respect to minima, and in harmonics that is with respect to all intervals less than concords. But by the prior reasons the aphelial movement not of Venus but of the Earth, will approximate the consonance of some number of octaves to be established with the aphelial movement of Saturn. For compound together, first, 4 : 5 the private ratio of Saturn's movements, i.e., from the aphelion to the perihelial of Saturn (Proposition XI), secondly, the 1 : 2 of the converging movements of Saturn and Jupiter, i.e., from the perihelion of Saturn to the aphelion of Jupiter (by Proposition VIII), thirdly, the 1 : 8 of the diverging movements of Jupiter and Mars, i.e., from the aphelion of Jupiter to the perihelion of Mars (by Proposition XIV), fourthly, the 2 : 3 of the converging movements of Mars and the Earth, i.e., from the perihelion of Mars to the aphelion of the Earth (by Proposition XV): you will find between the aphelion of Saturn and the perihelion of the Earth the compound ratio 1 : 30, which falls short of 1 : 32, or five octaves, by only 30 : 32, i.e., 15 : 16 or a semitone. And so, if a semitone, divided into particles smaller than the least concord, is compounded with these four elements there will be a perfect consonance of five octaves between the aphelial movements of Saturn and the Earth, which have been set forth. But in order for the same aphelial movement of Saturn to make some number of octaves with the aphelial movement of Venus, it would have been necessary to snatch approximately a whole perfect fourth from the prior reasons; for if you compound 3 : 5, which exists between the aphelial movements of the Earth and Venus, with the ratio 1 : 30 compounded of the four prior elements, then as it were from the prior reasons, 1 : 50 is found between the aphelial movements of Saturn and Venus: This interval differs from 1 : 32, or five octaves, by 32 : 50, i.e., 16 : 25, which is a perfect fifth and a diesis; and from six octaves, or 1 : 64, it differs by 50 : 64, i.e., 25 : 32, or a perfect fourth minus a diesis. Accordingly, an indentisonant consonance was due to be established, not between the aphelial movements of Venus and Saturn but between those of Venus and the

Earth, so that Saturn might keep a diversisonant consonance with Venus.

XXXII. PROPOSITION. *In the universal consonances of planets of the minor scale the exact aphelial movement of Saturn could not harmonize precisely with the other planets.*

For the Earth by its aphelial movement does not concur in the universal consonance of the minor scale, because the aphelial movements of the Earth and Venus make the interval 3 : 5, which is of the major scale (by Proposition XVII). But by its aphelial movement Saturn makes an identisonant consonance with the aphelial movement of the Earth (by Proposition XXXI). Therefore, neither does Saturn concur by its aphelial movement. Nevertheless, in place of the aphelial movement there follows some faster movement of Saturn, very near to the aphelial, and also in the minor scale—as was apparent in Chapter 7.

XXXIII. PROPOSITION. *The major kind of consonances and musical scale is akin to the aphelial movements; the minor to the perihelial.*

For although a major consonance (*dura harmonia*] is set up not only between the aphelial movement of the Earth and the aphelial movement of Venus but also between the lower aphelial movements and the lower movements of Venus as far as its perihelion; and, conversely, there is a minor consonance not merely between the perihelial movement of Venus and the perihelial of the Earth but also between the higher movements of Venus as far as the aphelion and the higher movements of the Earth (by Propositions XX and XXIV). Accordingly, the major scale is designated properly only in the aphelial movements, the minor, only in the perihelial.

XXXIV. PROPOSITION. *The major scale is more akin to the upper of the two planets, the minor, to the lower.*

[311] For, because the major scale is proper to the aphelial movements, the minor, to the perihelial (by the preceding proposition), while the aphelial are slower and graver than the perihelial; accordingly, the major scale is proper to the slower movements, the minor to the faster. But the upper of the two planets is more akin to the slow movements, the lower, to the fast, because slowness of the private movement always follows upon altitude in the world. Therefore, of two planets which adjust themselves to both modes, the upper is more akin to the major mode of the scale, the lower, to the minor. Further, the major scale employs the major intervals 4 : 5 and 3 : 5, and the minor, the minor ones, 5 : 6 and 5 : 8. But, moreover, the upper planet has both a greater sphere and slower, *i.e.*, greater movements and a lengthier circuit; but those things which agree greatly on both sides are rather closely united.

XXXV. PROPOSITION. *Saturn and the Earth embrace the major scale more closely Jupiter and Venus, the minor.*

For, first, the Earth, as compared with Venus and as designating both scales along with Venus, is the upper. Accordingly, by the preceding proposition, the Earth embraces the major scale chiefly; Venus, the minor. But with its aphelial movement Saturn harmonizes with the Earth's aphelial movement, through an octave (by Proposition XXXI): wherefore too (by Proposition XXXIII) Saturn embraces the major scale. Secondly, by the same proposition, Saturn by means of its aphelial movement nurtures more the major scale and (by Proposition XXXII) spits out the minor scale. Accordingly, it is more closely related to the major scale than to the minor, because the scales are properly designated by the extreme movements.

Now as regards Jupiter, in comparison with Saturn it is lower; therefore as the major scale is due to Saturn, so the minor is due to Jupiter, by the preceding proposition.

XXXVI. PROPOSITION. *The perihelial movement of Jupiter had to concord with the perihelial movement of Venus in one scale but not also in the same consonance; and all the less so, with the perihelial movement of the Earth.*

For, because the minor scale chiefly was due to Jupiter, by the preceding proposition, while the perihelial movements are more akin to the minor scale (by Proposition XXX), accordingly, by its perihelial movement Jupiter had to designate the key of the minor scale, *viz.*, its definite pitch or key-note [*phthongum*]. But too the perihelial movements of Venus and the Earth designate the same scale (by Proposition XXVIII); therefore the perihelial movement of Jupiter was to be associated with their perihelial movements in the same tuning, but it could not constitute a consonance with the perihelial movements of Venus. For, because (by Proposition VIII) it had to make about 1 : 3 with the aphelial movement of Saturn, *i.e.*, the note [*clavem*] d of that system, wherein the aphelial movement of Saturn strikes the note *G*, but the aphelial movement of Venus the note *e:* accordingly, it approached the note *e* within an interval of least consonance. For the least consonance is 5 : 6, but the interval between *d* and *e* is much smaller, *viz.*, 9 : 10, a whole tone. And although in the perihelial tension [*tensione*] Venus is raised from the *d* of the aphelial tension yet this elevation is less than a diesis, (by Proposition XXVIII). But the diesis (and hence any smaller interval) if compounded with a minor whole tone does not yet equal 5 : 6 the interval of least consonance. Accordingly, the perihelial movement of Jupiter could not observe 1 : 3 or thereabouts with the aphelial movement of Saturn and at the same time harmonize with Venus. Nor with the Earth. For if the perihelial movement of Jupiter had been adjusted to the key of the perihelial movement of Venus in the same tension in such fashion that below the quantity of least concord it should preserve with the aphelial movement of Saturn the interval 1 : 3, *viz.*, by differing from the perihelial movement of Venus by a minor whole tone, 9 : 10 or 36 : 40 (besides some octaves) towards the low. Now the perihelial movement of the Earth differs from the same perihelial movement of Venus by 5 : 8, *i.e.*, by 25 : 40. And so the perihelial movements of the Earth and Jupiter differ by 25 : 36, over and above some number of octaves. But that is not harmonic, because it is the square of 5 : 6, or a perfect fifth diminished by one diesis.

XXXVII. PROPOSITION. *It was necessary for an interval equal to the interval of Venus to accede to the 2 : 3 of the compounded private consonances of Saturn and Jupiter and to 1 : 3 the great consonance common to them.*

For with its aphelial movement Venus assists in the proper designation of the major scale; with its perihelial, that of the minor scale, by Propositions XXVII and XXXIII. But by its aphelial movement Saturn had to be in concord also with the major scale and thus with the aphelial movement of Venus, by Proposition XXXV, but Jupiter's perihelial with the perihelial of Venus, by the preceding proposition. Accordingly, as great as Venus makes its interval from aphelial to perihelial to be, so great an interval must also accede to that movement of Jupiter which makes 1 : 3 with the aphelial movement of Saturn—to the very perihelial movement of Jupiter. But the consonance of the converging movements of Jupiter and Saturn is precisely 1 : 2, by Proposition VIII. Accordingly, if the interval 1 : 2 is divided into the interval [312] greater than 1 : 3, there results, as the compound of the private ratios of both, something which is proportionately greater than 2 : 3.

Above, in Proposition xxvi, the private ratio of the movements of Venus was 243 : 250 or approximately 35 : 36; but in Chapter 4, between the aphelial movement of Saturn and the perihelial movement of Jupiter there was found a slightly greater excess beyond 1 : 3, *viz.*, between 26 : 27 and 27 : 28. But the quantity here prescribed is absolutely equalled, by the addition of a single second to the aphelial movement of Saturn, and I do not know whether astronomy can discern that difference.

XXXVIII. PROPOSITION. *The increment 243 : 250 to 2 : 3, the compound of the private ratios of Saturn and Jupiter, which was up to now being established by the prior reasons, was to be distributed among the planets in such fashion that of it the comma 80 : 81 should accede to Saturn and the remainder, 19,683 : 20,000 or approximately 62 : 63, to Jupiter.*

It follows from Axiom xix that this was to have been distributed between both planets so that each could with some latitude concur in the universal consonances of the scale akin to itself. But the interval 243 : 250 is smaller than all concords: accordingly no harmonic rules remain whereby it may be divided into two concordant parts, with the single exception of those of which there was need in the division of 24 : 25, the diesis, above in Proposition xxvi; namely, in order that it may be divided into the comma 80 : 81 (which is a primary one of those intervals which are subordinate to the concordant) and into the remainder 19,683 : 20,000, which is slightly greater than a comma, *viz.*, approximately 62 : 63. But not two but one comma had to be taken away, lest the parts should become too unequal, since the private ratios of Saturn and Jupiter are approximately equal (according to Axiom x extended even to concords and parts smaller than those) and also because the comma is determined by the intervals of the major whole tone and minor whole tone, not so two commas. Furthermore, to Saturn the higher and mightier planet was due not that part which was greater, although Saturn had the greater private consonance 4 : 5, but that one which is prior and more beautiful, *i.e.*, more harmonic. For in Axiom x the consideration of priority and harmonic perfection comes first, and the consideration of quantity comes last, because there is no beauty in quantity of itself. Thus the movements of Saturn become 64 : 81, an adulterine[1] major third, as we have called them in Book iii, Chapter 12, but those of Jupiter, 6,561 : 8,000.

I do not know whether it should be numbered among the causes of the addition of a comma to Saturn that the extreme intervals of Saturn can constitute the ratio 8 : 9, the major whole tone, or whether that resulted without further ado from the preceding causes of the movements. Accordingly, you here have, in place of a corollary, the reason why, above in Chapter 4, the intervals of Saturn were found to embrace approximately a major whole tone.

XXXIX. PROPOSITION. *Saturn could not harmonize with its exact perihelial movement in the universal consonances of the planets of the major scale, nor Jupiter with its exact aphelial movement.*

For since the aphelial movement of Saturn had to harmonize exactly with the aphelial movements of the Earth and Venus (by Proposition xxxi), that movement of Saturn which is 4 : 5 or one major third faster than its aphelial will also

[1]See footnote to *Intervals Compared with Harmonic Ratios*, p. 186.

harmonize with them. For the aphelial movements of the Earth and Venus make a major sixth, which, by the demonstrations of Book III, is divisible into a perfect fourth and a major third, therefore the movement of Saturn, which is still faster than this movement already harmonized but none the less below the magnitude of a concordant interval, will not exactly harmonize. But such a movement is Saturn's perihelial movement itself, because it differs from its aphelial movement by more than the interval 4 : 5, *viz.*, one comma or 80 : 81 more (which is less than the least concord), by Proposition XXXVIII. Accordingly the perihelial movement of Saturn does not exactly harmonize. But neither does the aphelial movement of Jupiter do so precisely. For while it does not harmonize precisely with the perihelial movement of Saturn, it harmonizes at a distance of a perfect octave (by Proposition VIII), wherefore, according to what has been said in Book III, it cannot precisely harmonize.

XL. PROPOSITION. *It was necessary to add the lemma of Plato to 1 : 8, or the triple octave, the joint consonance of the diverging movements of Jupiter and Mars established by the prior reasons.*

For because, by Proposition XXXI, there had to be 1 : 32, *i.e.*, 12 : 384, between the aphelial movements of Saturn and the Earth, but there had to be 3 : 2, *i.e.*, 384 : 256, from the aphelion of the Earth to the perihelion of Mars [313] (by Proposition XV), and from the aphelion of Saturn to its perihelion, 4 : 5 or 12 : 15 with its increment (by Proposition XXXVIII); finally, from the perihelion of Saturn to the aphelion of Jupiter 1 : 2 or 15 : 30 (by Proposition VIII); accordingly, there remains 30 : 256 from the aphelion of Jupiter to the perihelion of Mars, by the subtraction of the increment of Saturn. But 30 : 256 exceeds 32 : 256 by the interval 30 : 32, *i.e.*, 15 : 16 or 240 : 256, which is a semitone. Accordingly, if the increment of Saturn, which (by Proposition XXXVIII) had to be 80 : 81, *i.e.*, 240 : 243, is compounded inversely with 240 : 243, the result is 243 : 256; but that is the lemma of Plato,[1] *viz.*, approximately 19 : 20, see Book III. Accordingly, Plato's lemma had to be compounded with the 1 : 8.

And so the great ratio of Jupiter and Mars, *viz.*, of the diverging movements, ought to be 243 : 2,048, which is somehow a mean between 243 : 2,187 and 243 : 1,944, *i.e.*, between 1 : 9 and 1 : 8, whereof proportionality required the first, above; and a nearer harmonic concord, the second.

XLI. PROPOSITION. *The private ratio of the movements of Mars has necessarily been made the square of the harmonic ratio 5 : 6, viz., 25 : 36.*

For, because the ratio of the diverging movements of Jupiter and Mars had to be 243 : 2,048, *i.e.*, 729 : 6,144, by the preceding proposition, but that of the converging movements 5 : 24, *i.e.*, 1,280 : 6,144 (by Proposition XIII), therefore the compound of the private ratios of both was necessarily 729 : 1,280 or 72,900 : 128,000. But the private ratio of Jupiter alone had to be 6,561 : 8,000, *i.e.*, 104,976 : 128,000 (by Proposition XXVIII). Therefore, if the compound ratio of both is divided by this, the private ratio of Mars will be left as 72,900 : 104,976, *i.e.*, 25 : 36, the square root of which is 5 : 6.

In another fashion, as follows: There is 1 : 32 or 120 : 3,840 from the aphelial movement of Saturn to the aphelial movement of the Earth, but from that same movement to the perihelial of Jupiter there is 1 : 3 or 120 : 360, with its increment. But from this to the aphelial movement of Mars is 5 : 24 or 360 : 1,728. Accordingly, from the aphelial movement of Mars to the aphelial move-

[1] *Timaeus*, 36.

ment of the Earth, there remains 1,728 : 3,840 minus the increment of the ratio
of the diverging movements of Saturn and Jupiter. But from the same aphelial
movement of the Earth to the perihelial of Mars there is 3 : 2, *i.e.*, 3,840 : 2,500.
Therefore between the aphelial and perihelial movements of Mars there remains
the ratio 1,728 : 2,560, *i.e.*, 27 : 40 or 81 : 120, minus the said increment. But
81 : 120 is a comma less than 80 : 120 or 2 : 3. Therefore, if a comma is taken
away from 2 : 3, and the said increment (which by Proposition XXXVIII is equal
to the private ratio of Venus) is taken away too, the private ratio of Mars is
left. But the private ratio of Venus is the diesis diminished by a comma, by
Proposition XXVI. But the comma and the diesis diminished by a comma make
a full diesis or 24 : 25. Therefore if you divide 2 : 3, *i.e.*, 24 : 36 by the diesis
24 : 25, Mars' private ratio of 25 : 36 is left, as before, the square root of which,
or 5 : 6, goes to the intervals, by Chapter 3.

Behold again the reason why—above, in Chapter 4—the extreme intervals of
Mars have been found to embrace the harmonic ratio 5 : 6.

XLII. PROPOSITION. *The great ratio of Mars and the Earth, or the common ratio
of the diverging movements, has been necessarily made to be 54 : 125, smaller than
the consonance 5 : 12 established by the prior reasons.*

For the private ratio of Mars had to be a perfect fifth, from which a diesis
has been taken away, by the preceding proposition. But the common or minor
ratio of the converging movements of Mars and the Earth had to be a perfect
fifth or 2 : 3, by Proposition XV. Finally, the private ratio of the Earth is the
diesis squared, from which a comma is taken away, by Propositions XXVI and
XXVIII. But out of these elements is compounded the major ratio or that of the
diverging movements of Mars and the Earth—and it is two perfect fifths (or
4 : 9, *i.e.*, 108 : 243) plus a diesis diminished by a comma, *i.e.*, plus 243 : 250;
namely, it is 108 : 250 or 54 : 125, *i.e.*, 608 : 1,500. But this is smaller than
625 : 1,500, *i.e.*, than 5 : 12, in the ratio 602 : 625, which is approximately 36 : 37,
smaller than 625 : 1,500, *i.e.*, than 5 : 12, in the ratio 602 : 625, which is approximately 36 : 37, smaller than the least concord.

XLIII. PROPOSITION. *The aphelial movement of Mars could not harmonize in
some universal consonance; nevertheless it was necessary for it to be in concord to
some extent in the scale of the minor mode.*

For, because the perihelial movement of Jupiter has the pitch *d* of acute
tuning in the minor mode, and the consonance 5 : 24 ought to have existed
between that and the aphelial movement of Mars, therefore, the aphelial movement of Mars occupies the adulterine pitch of the same acute tuning. I say
adulterine for, although in Book III, Chapter 12, the adulterine consonances
were reviewed and deduced from the composition of systems, certain ones
which exist in the simple natural system were omitted. [314] And so, after the
line which ends 81 : 120, the reader may add: if you divide into it 4 : 5 or 32 : 40,
there remains 27 : 32, the subminor sixth,[1] which exists between *d* and *f* or *c* and *e²*
or *a* and *c* of even the simple octave. And in the ensuing table, the following
should be in the first line; for 5 : 6 there is 27 : 32, which is deficient.

From that it is clear that in the natural system the true note [*clavem*] *f*, as
regulated by my principles, constitutes a deficient or adulterine minor sixth with
the note *d*. Accordingly since between the perihelial movement of Jupiter set

[1]Here "sixth" (*sexta*) should probably be "third" (*tertia*). E. C., Jr.
[2]*C* and *e* do not produce a subminor third in the "natural system." E. C., Jr.

up in the true note d and the aphelial movement of Mars there is a perfect minor sixth over and above the double octave, but not the diminished (by Proposition XIII), it follows that with its aphelial movement Mars designates the pitch which is one comma higher than the true note f; and so it will concord not absolutely but merely to a certain extent in this scale. But it does not enter into either the pure or the adulterine universal harmony. For the perihelial movement of Venus occupies the pitch of e in this tuning [tensionem]. But there is dissonance between e and f, on account of their nearness. Therefore, Mars is in discord with the perihelial movement of one of the planets, viz., Venus. But too it is in discord with the other movements of Venus; they are diminished by a comma less than a diesis: wherefore, since there is a semitone and a comma between the perihelial movement of Venus and the aphelial movement of Mercury, accordingly, between the aphelion of Venus and the aphelion of Mars there will be a semitone and a diesis (neglecting the octaves), i.e., a minor whole tone, which is still a dissonant interval. Now the aphelial movement of Mars concords to that extent in the scale of the minor mode, but not in that of the major. For since the aphelial movement of Venus concords with the e of the major mode, while the aphelial movement of Mars (neglecting the octaves) has been made a minor whole tone higher than e, then necessarily the aphelial movement of Mars in this tuning would fall midway between f and f sharp and would make with g (which in this tuning would be occupied by the aphelial movement of the Earth) the plainly discordant interval 25 : 27, viz., a major whole tone diminished by a diesis.

In the same way, it will be proved that the aphelial movement of Mars is also in discord with the movements of the Earth. For because it makes a semitone and comma with the perihelial movement of Venus, i.e., 14 : 15 (by what has been said), but the perihelial movements of the Earth and Venus make a minor sixth 5 : 8 or 15 : 24 (by Proposition XXVII). Accordingly, the aphelial movement of Mars together with the perihelial movement of the Earth (the octaves added to it) will make 14 : 24 or 7 : 12, a discordant interval and one not harmonic, like 7 : 6. For any interval between 5 : 6 and 8 : 9 is dissonant and discordant, as 6 : 7 in this case. But no other movement of the Earth can harmonize with the aphelial movement of Mars. For it was said above that it makes the discordant interval 25 : 27 with the Earth (neglecting the octaves); but all from 6 : 7 or 24 : 28 to 25 : 27 are smaller than the least harmonic interval.

XLIV. COROLLARY. *Accordingly it is clear from the above Proposition XLIII concerning Jupiter and Mars, and from Proposition XXXIX concerning Saturn and Jupiter, and from Proposition XXXVI concerning Jupiter and the Earth, and from Proposition XXXII concerning Saturn, why—in Chapter 5, above—it was found that all the extreme movements of the planets had not been adjusted perfectly to one natural system or musical scale, and that all those which had been adjusted to a system of the same tuning did not distinguish the pitches [loca] of that system in a natural way or effect a purely natural succession of concordant intervals. For the reasons are prior whereby the single planets came into possession of their single consonances; those whereby all the planets, of the universal consonances; and finally, those whereby the universal consonances of the two modes, the major and the minor: when all those have been posited, an omniform adjustment to one natural system is prevented. But if those causes had not necessarily come first, there is no doubt that either one system and one tuning of it would have embraced the extreme movements of all the planets; or, if there was need of two systems for the two modes of song, the major and minor, the very order of the natural scale would have been expressed not merely in one mode, the major, but also in the remaining minor mode. Accordingly, here in Chapter 5, you have the promised causes of the discords through least intervals and intervals smaller than all concords.*

XLV. PROPOSITION. *It was necessary for an interval equal to the interval of Venus to be added to the common major consonance of Venus and Mercury, the double octave, and also the private consonance of Mercury, which were established above in Propositions XII and XIII by the prior reasons,* [315] *in order that the private ratio of Mercury should be a perfect 5 : 12 and that thus Mercury should with both its movements harmonize with the single perihelial movement of Venus.*

For, because the aphelial movement of Saturn, the highest and outmost planet, circumscribed around its regular solid, had to harmonize with the aphelial movement of the Earth, the highest movement of the Earth, which divides the classes of figures; it follows by the laws of opposites that the perihelial movement of Mercury as the innermost planet, inscribed in its figure, the lowest and nearest to the sun, should harmonize with the perihelial movement of the Earth, with the lowest movement of the Earth, the common boundary: the former in order to designate the major mode of consonances, the latter the minor mode, by Propositions XXXIII and XXXIV. But the perihelial movement of Venus had to harmonize with the perihelial movement of the Earth in the consonance 5 : 3, by Proposition XXVII; therefore too the perihelial movement of Mercury had to be tempered with the perihelial of Venus in one scale. But by Proposition XII the consonance of the diverging movements of Venus and Mercury was determined by the prior reasons to be 1 : 4; therefore, now by these posterior reasons it was to be adjusted by the accession of the total interval of Venus. Accordingly, not from further on, from the aphelion, but from the perihelion of Venus to the perihelion of Mercury there is a perfect double octave. But the consonance 3 : 5 of the converging movements is perfect, by Proposition XV. Accordingly if 1 : 4 is divided by 3 : 5, there remains to Mercury singly the private ratio 5 : 12, perfect too, but not further (by Proposition XVI, through the prior reasons) diminished by the private ratio of Venus.

Another reason. Just as only Saturn and Jupiter are touched nowhere on the outside by the dodecahedron and icosahedron wedded together, so only Mercury is untouched on the inside by these same solids, since they touch Mars on the inside, the Earth on both sides, and Venus on the outside. Accordingly, just as something equal to the private ratio of Venus has been added distributively to the private ratios of movements of Saturn and Jupiter, which are supported by the cube and tetrahedron; so now something as great was due to accede to the private ratio of solitary Mercury, which is comprehended by the associated figures of the cube and tetrahedron; because, as the octahedron, a single figure among the secondary figures, does the job of two among the primary, the cube and tetrahedron (concerning which see Chapter 1), so too among the lower planets there is one Mercury in place of two of the upper planets, *viz.*, Saturn and Jupiter.

Thirdly, just as the aphelial movement of the highest planet Saturn had to harmonize, in some number of octaves, *i.e.*, in the continued double ratio, 1 : 32, with the aphelial movement of the higher and nearer of the two planets which shift the mode of consonance (by Proposition XXXI); so, *vice versa*, the perihelial movement of the lowest planet Mercury, again through some number of octaves, *i.e.*, in the continued double ratio, 1 : 4, had to harmonize with the perihelial movement of the lower and similarly nearer of the two planets which shift the mode of consonance.

Fourthly, of the three upper planets, Saturn, Jupiter, and Mars, the single but extreme movements concord with the universal consonances; accordingly both extreme movements of the single lower planet, *viz.*, Mercury, had to concord with the same; for the middle planets, the Earth and Venus, had to shift the mode of consonances, by Propositions XXXIII and XXXIV.

Finally, in the three pairs of the upper planets perfect consonances have been found between the converging movements, but adjusted [*fermentatae*] consonances between the diverging movements and private ratios of the single planets; accordingly, in the two pairs of the lower planets, conversely, perfect consonances had to be found not between the converging movements chiefly, nor between the diverging, but between the movements of the same field. And because two perfect consonances were due to the Earth and Venus, therefore two perfect consonances were due to Venus and Mercury also. And the Earth and Venus had to receive as perfect a consonance between their aphelial movements as between their perihelial, because they had to shift the mode of their consonance; but Venus and Mercury, as not shifting the mode of their consonance, did not also require perfect consonances between both pairs, the aphelial movements and the perihelial; but there came in place of the perfect consonance of the aphelial movements, as being already adjusted the perfect consonance of the converging movements, so that just as Venus, the higher of the lower planets, has the least private ratio of all the private ratios of movements (by Proposition XXVI), and Mercury, the lower of the lower, has received the greatest ratio of all the private ratios of movements (by Proposition XXX), so too the private ratio of Venus should be the most imperfect of all the private ratios or the farthest removed from consonances, while the private ratio of Mercury should be most perfect of all the private ratios, *i.e.*, an absolute consonance without adjustment, and that finally the relations should be everywhere opposite.

For He Who is before the ages and on into the ages thus adorned the great things of His wisdom: nothing excessive, nothing defective, no room for any censure. How lovely are his works! All things, in twos, one [316] against one, none lacking its opposite. He has strengthened the goods—adornment and propriety—of each and every one and established them in the best reasons, and who will be satiated seeing their glory?

XLVI. AXIOM. *If the interspacing of the solid figures between the planetary spheres is free and unhindered by the necessities of antecedent causes, then it ought to follow to perfection the proportionality of geometrical inscriptions and circumscriptions, and thereby the conditions of the ratio of the inscribed to the circumscribed spheres.*

For nothing is more reasonable than that physical inscription should exactly represent the geometrical, as the work, its pattern.

XLVII. PROPOSITION. *If the inscription of the regular solids among the planets was free, the tetrahedron was due to touch with its angles precisely the perihelial sphere of Jupiter above it, and with centres of its planes precisely the aphelial sphere of Mars below it. But the cube and the octahedron, each placing its angles in the perihelial sphere of the planet above, were due to penetrate the sphere of the inside planet with the centres of their planes, in such fashion that those centres should turn within the aphelial and perihelial spheres: on the other hand, the dodecahedron and icosahedron, grazing with their angles the perihelial spheres of their planets on the outside, were due not quite to touch with the centres of their planes the aphelial spheres of their inner planets. Finally, the dodecahedral echinus, placing its angles in the perihelial sphere of Mars, was due to come very close to the aphelial sphere of Venus with the midpoints of its converted sides which interdistinguish two solid rays.*

For the tetrahedron is the middle one of the primary figures, both in genesis and in situation in the world; accordingly, it was due to remove equally both regions, that of Jupiter and that of Mars. And because the cube was above it and outside it, and the dodecahedron was below it and within it, therefore it was natural that their inscription should strive for the contrariety wherein the tetrahedron held a mean, and that the one of them should make an excessive inscription, and the other a defective, *viz.*, the one should somewhat penetrate the inner sphere, the other not touch it. And because the octahedron is cognate to the cube and has an equal ratio of spheres, but the icosahedron to the dodecahedron, accordingly, whatever the cube has of perfection of inscription, the same was due to the octahedron also, and whatever the dodecahedron, the same to the icosahedron too. And the situation of the octahedron's similar to the situation of the cube, but that of the icosahedron to the situation of the dodecahedron, because as the cube occupies the one limit to the outside, so the octahedron occupies the remaining limit to the inside of the world, but the dodecahedron and icosahedron are midway: accordingly even a similar inscription was proper, in the case of the dodecahedron, one penetrating the sphere of the inner planet, in that of the icosahedron, one falling short of it.

But the echinus, which represents the icosahedron with the apexes of its angles and the dodecahedron with the bases, was due to fill, embrace, or dispose both regions, that between Mars and the Earth with the dodecahedron as well as that between the Earth and Venus with the icosahedron. But the preceding axiom makes clear which of the opposites was due to which association. For the tetrahedron, which has a rational inscribed sphere, has been allotted the middle position among the primary figures and is surrounded on both sides by figures of incommensurable spheres, whereof the outer is the cube, the inner the dodecahedron, by Chapter 1 of this book. But this geometrical quality, *viz.*, the rationality of the inscribed sphere, represents in nature the perfect inscription of the planetary sphere. Accordingly, the cube and its allied figure have their inscribed spheres rational only in square, *i.e.*, in power alone; accordingly, they ought to represent a semiperfect inscription, where, even if not the extremity of the planetary sphere, yet at least something on the inside and rightfully a mean between the aphelial and perihelial spheres—if that is possible through other reasons—is touched by the centres of the planes of the figures. On the other hand, the dodecahedron and its allied figure have their inscribed spheres clearly irrational both in the length of the radius and in the square; accordingly, they ought to represent a clearly imperfect inscription and one touching absolutely nothing of the planetary sphere, *i.e.*, falling short and not reaching as far as the aphelial sphere of the planet with the centres of its planes.

Although the echinus is cognate to the dodecahedron and its allied figure, nevertheless it has a property similar to the tetrahedron. For the radius of the sphere inscribed in its inverted sides is indeed incommensurable with the radius of the circumscribed sphere, but it is, however, commensurable with the length of the distance between two neighbouring angles. And so the perfection of the commensurability of rays is approximately as great as in the tetrahedron; but elsewhere the imperfection is as great as in the [317] dodecahedron and its allied figure. Accordingly it is reasonable too that the physical inscription belonging to it should be neither absolutely tetrahedral nor absolutely dodecahedral but of an intermediate kind; in order that (because the tetrahedron was due to touch the extremity of the sphere with its planes, and the dodecahedron, to fall short of it by a definite interval) this wedge-shaped figure with the inverted sides should stand between the icosahedral space and the extremity of the inscribed sphere and should nearly touch this extremity—if nevertheless this figure was to be admitted into association with the remaining five, and if its laws could be allowed, with the laws of the others remaining. Nay, why do I say "could be allowed"? For they could not do without them. For if an inscription, which was loose and did not come into contact fitted the dodecahedron, what else could confine that indefinite looseness within the limits of a fixed magnitude, except this subsidiary figure cognate to the dodecahedron and icosahedron, and which comes almost into contact with its inscribed sphere and does not fall short (if indeed it does fall short) any more than the tetrahedron exceeds and penetrates —with which magnitude we shall deal in the following.

This reason for the association of the echinus with the two cognate figures (*viz.*, in order that the ratio of the spheres of Mars and Venus, which they had left indefinite, should be made determinate) is rendered very probable by the fact that 1,000, the semidiameter of the sphere of the Earth, is found to be practically a mean proportional between the perihelial sphere of Mars and the aphelial sphere of Venus; as if the interval, which the echinus assigns to the cognate figures, has been divided between them as proportionally as possible.

XLVIII. PROPOSITION. *The inscription of the regular solid figures between the planetary spheres was not the work of pure freedom; for with respect to very small magnitudes it was hindered by the consonances established between the extreme movements.*

For, by Axioms I and II, the ratio of the spheres of each figure was not due to be expressed immediately by itself, but by means of it the consonances most akin to the ratios of the spheres were first to be sought and adjusted to the extreme movements.

Then, in order that, by Axioms XVIII and XX, the universal consonances of the two modes could exist, it was necessary for the greater consonances of the single pairs to be readjusted somewhat, by means of the posterior reasons. Accordingly, in order that those things might stand, and be maintained by their own reasons, intervals were required which are somewhat discordant with those which arise from the perfect inscription of figures between the spheres, by the laws of movements unfolded in Chapter 3. In order that it be proved and made manifest how much is taken away from the single planets by the consonances established by their proper reasons; come, let us build up, out of them, the intervals of the planets from the sun, by a new form of calculation not previously tried by anyone.

Now there will be three heads to this inquiry: First, from the two extreme movements of each planet the similar extreme intervals between it and the sun will be investigated, and by means of them the radius of the sphere in those di-

234 KEPLER

mensions, of the extreme intervals, which are proper to each planet. Secondly, by means of the same extreme movements, in the same dimensions for all, the mean movements and their ratio will be investigated. Thirdly, by means of the ratio of the mean movements already disclosed, the ratio of the spheres or mean intervals and also one ratio of the extreme intervals, will be investigated; and the ratio of the mean intervals will be compared with the ratios of the figures.

As regards the first: we must repeat, from Chapter 3, Article VI, that the ratio of the extreme movements is the inverse square of the ratio of the corresponding intervals from the sun. Accordingly, since the ratio of the squares is the square of the ratio of its sides, therefore, the numbers, whereby the extreme movements of the single planets are expressed, will be considered as squares and the extraction of their roots will give the extreme intervals, whereof it is easy to take the arithmetic mean as the semidiameter of the sphere and the eccentricity. Accordingly the consonances so far established have prescribed.

[318]Planets Props.	Ratios of movements	The roots either prolonged or of their multiples	Therefore the semidiameter of the sphere	Eccen-tricity	In dimensions whereof the semidiameter of the sphere is 100,000
Saturn by XXXVIII	64 : 81	80 : 90	85	5	5,882
Jupiter by XXXVIII	6,561 : 8,000	81,000 : 89,444	85,222	4,222	4,954
Mars by XLI	25 : 36	50 : 60	55	5	9,091
Earth by XXVIII	2,916 : 3,125	93,531 : 96,825	95,178	1,647	1,730
Venus by XXVIII	243 : 250	9,859 : 10,000	99,295	705	710
Mercury by XLV	5 : 12	63,250 : 98,000	80,625	17,375	21,551

For the second of the things proposed, we again have need of Chapter 3, Article XII, where it was shown that the number which expresses the movement which is as a mean in the ratio of the extremes is less than their arithmetic mean, also less than the geometric mean by half the difference between the geometric and arithmetic means. And because we are investigating all the mean movements in the same dimensions, therefore let all the ratios hitherto established between different twos and also all the private ratios of the single planets be set out in the measure of the least common divisible. Then let the means be sought: the arithmetic, by taking half the difference between the extreme movements of each planet, the geometric, by the multiplication of one extreme into the other and extracting the square root of the product; then by subtracting half the difference of the means from the geometric mean, let the number of the mean movement be constituted in the private dimensions of each planet, which can easily, by the rule of ratios, be converted into the common dimensions.

[319] Therefore, from the prescribed consonances, the ratio of the mean diurnal movements has been found, viz., the ratio between the numbers of the degrees and minutes of each planet. It is easy to explore how closely that approaches to astronomy.

Harmonic ratios of two				*Numbers of the extreme movements*	*Private ratios of the single planets*	*Continued means of the single planets*		*Halves of the difference*	*Number of the mean movement in dimensions*	
						Arithmetic	*Geometric*		*Private*	*Common*
1			♄	139,968	64	72.50	72.00	.25	71.75	156,917
	1		♄	177,147	81					
	2		♃	354,294	6,561	7,280.5	7,244.9	17.8	7,227.1	390,263
	5		♃	432,000	8,000					
	24		♂	2,073,600	25	30.50	30.00	.25	29.75	2,467,584
	2		♂	2,985,984	36					
32	3		♁	4,478,976	2,916	3,020.500	3,018.692	.904	3,017.788	4,635,322
		5	♁	4,800,000	3,125					
	5	8	♀	7,464,960	243	246.500	246.475	.0125	246.4625	7,571,328
1	3		♀	7,680,000	250					
	5		☿	12,800,000	5	8.500	7.746	.377	7.369	18,864,680
4			☿	30,720,000	12					

The third head of things proposed requires Chapter 3, Article VIII. For when the ratio of the mean diurnal movements of the single planets has been found, it is possible to find the ratio of the spheres too. For the ratio of the mean movements is the $3/2$th power of the inverse ratio of the spheres. But, too, the ratio of the cube numbers is the $3/2$th power of the ratio of the squares of those same square roots, given in the table of Clavius, which he subjoined to his *Practical Geometry*. Wherefore, if the numbers of our mean movements (curtailed, if need be, of an equal number of ciphers) are sought among the cube numbers of that table, they will indicate on the left, under the heading of the squares, the numbers of the ratio of the spheres; then the eccentricities ascribed above to the single planets in the private ratio of the semidiameters of each may easily be converted by the rule of ratios into dimensions common to all, so that, by their addition to the semidiameters of the spheres and subtraction from them, the extreme intervals of the single planets from the sun may be established. Now we shall give to the semidiameter of the terrestrial sphere the round number 100,000, as is the practice in astronomy, and with the following design: because this number or its square or its cube is always made up of mere ciphers; and so too we shall raise the mean movement of the Earth to the number 10,000,000,000 and by the rule of ratios make the number of the mean movement of any planet be to the number of the mean movement of the Earth, as 10,000,000,000 is to the new measurement. And so the business can be carried on with only five cube roots, by comparing those single cube roots with the one number of the Earth.

Numbers of the mean movements		Numbers of the ratio of the spheres found among the squares	Semi-diameters as above	Eccentricities in dimensions		Extreme intervals resulting	
In the original dimensions	In the new dimensions found in inverse order among the cubes			Private as above	Common	Aphelion	Perihelion
♄ 156,917	29,539,960	9,556	85	5	562	10,118	8,994
♃ 390,263	11,877,400	5,206	85,222	4,222	258	5,464	4,948
♂ 2,467,584	1,878,483	1,523	55	5	138	1,661	1,384
♁ 4,635,322	1,000,000	1,000	95,178	1,647	17	1,017	983
♀ 7,571,328	612,220	721	99,295	705	5	726	716
☿ 18,864,680	245,714	392	80,625	17,375	85	476	308

Accordingly, it is apparent in the last column what the numbers turn out to be whereby the converging intervals of two planets are expressed. All of them approach very near to those intervals, which I found from Brahe's observations. In Mercury alone is there some small difference. For astronomy is seen to give the following intervals to it: 470, 388, 306, all shorter. It seems that the reason for the dissonance may be referred either to the fewness of the observations or to the magnitude of the eccentricity. (See Chapter 3). But I hurry on to the end of the calculation.

For now it is easy to compare the ratio of the spheres of the figures with the ratio of the converging intervals.

[320] For if the semidiameter of the sphere circumscribed around the figure

which is commonly 100,000	becomes:	Then the semidiameter of the sphere or circle inscribed in:	instead of:	becomes:	Although by the consonances the interval is:		
In the cube	8,994	♄ [Saturn]	57,735	5,194	Mean	♃	5,206
In the tetrahedron	4,948	♃ [Jupiter]	33,333	1,649	Aphelial	♂	1,661
In the dodecahedron	1,384	♂ [Mars]	79,465	1,100	Aphelial	♁	1,018
In the icosahedron	983	♁ [Earth]	79,465	781	Aphelial	♀	726
In the echinus	1,384	♂ [Mars]	52,573	728	Aphelial	♀	726
In the octahedron	716	♀ [Venus	57,735	413	Mean	☿	392
In the square in the octahedron	716	♀ [Venus]	70,711	506	Aphelial	☿	476
	or 476	☿ [Mercury]	70,711	336	Perihelial	☿	308

That is to say, the planes of the cube extend down slightly below the middle circle of Jupiter; the octahedral planes, not quite to the middle circle of Mercury; the tetrahedral, slightly below the highest circle of Mars; the sides of the echinus, not quite to the highest circle of Venus; but the planes of the dodecahedron fall far short of the aphelial circle of the Earth; the planes of the icosa-dron also fall short of the aphelial circle of Venus, and approximately propor-tionally; finally, the square in the octahedron is quite inept, and not unjustly, for what are plane figures doing among solids? Accordingly, you see that if the planetary intervals are deduced from the harmonic ratios of movements hither-to demonstrated, it is necessary that they turn out as great as these allow, but not as great as the laws of free inscription prescribed in Proposition XLV would require: because this κόσμος γεωμέτρικος [geometrical adornment] of perfect in-scription was not fully in accordance with that other κόσμον ἁρμόνικον ἐνδεχόμενον [possible harmonic adornment]—to use the words of Galen, taken from the epigraph to this Book v. So much was to be demonstrated by the calculation of numbers, for the elucidation of the prescribed proposition.

I do not hide that if I increase the consonance of the diverging movements of Venus and Mercury by the private ratio of the movements of Venus, and, as a consequence, diminish the private ratio of Mercury by the same, then by this process I produce the following intervals between Mercury and the sun: 469, 388,307, which are very precisely represented by astronomy. But, in the first place, I cannot defend that diminishing by harmonic reasons. For the aphelial movement of Mercury will not square with that musical scale, nor in the planets which are opposite in the world is the planetary principle [ratio] of opposition of all conditions kept. Finally, the mean diurnal movement of Mercury becomes too great, and thereby the periodic time, which is the most certain fact in all astronomy, is shortened too much. And so I stay within the harmonic polity here employed and confirmed throughout the whole of Chapter 9. But none the less with this example I call you all forth, as many of you as have happened to read this book and are steeped in the mathematical disciplines and the knowl-edge of highest philosophy: work hard and either pluck up one of the conso-nances applied everywhere, interchange it with some other, and test whether or not you will come so near to the astronomy posited in Chapter 4, or else try by reasons whether or not you can build with the celestial movements something better and more expedient and destroy in part or in whole the layout applied by me. But let whatever pertains to the glory of Our Lord and Founder be equally permissible to you by way of this book, and up to this very hour I myself have taken the liberty of everywhere changing those things which I was able to discover on earlier days and which were the conceptions of a sluggish care or hurrying ardour.

[321] XLIX. ENVOI. *It was good that in the genesis of the intervals the solid figures should yield to the harmonic ratios, and the major consonances of two planets to the universal consonances of all, in so far as this was necessary.*

With good fortune we have arrived at 49, the square of 7; so that this may come as a kind of Sabbath, since the six solid eights of discourse concerning the construction of the heavens has gone before. Moreover, I have rightly made an *envoi* which could be placed first among the axioms: because God also, enjoying the works of His creation, "saw all things which He had made, and behold! they were very good."

There are two branches to the *envoi:* First, there is a demonstration concern-ing consonances in general, as follows: For where there is choice among different things which are not of equal weight, there the more excellent are to be put first and the more vile are to be detracted from, in so far as that is necessary, as the very word ὁ κόσμος, which signifies *adornment*, seems to argue. But inasmuch as life is more excellent than the body, the form than the material, by so much does harmonic adornment excel the geometrical.

For as life perfects the bodies of animate things, because they have been born for the exercise of life—as follows from the archetype of the world, which is the divine essence—so movement measures the regions assigned to the planets, each that of its own planet: because that region was assigned to the planet in order that it should move. But the five regular solids, by their very name, pertain to the intervals of the regions and to the number of them and the bodies; but the consonances to the movements. Again, as matter is diffuse and indefinite of itself, the form definite, unified, and determinant of the material, so too there are an infinite number of geometric ratios, but few consonances. For although among the geometrical ratios there are definite degrees of determinations, formation, and restriction, and no more than three can exist from the ascription of spheres to the regular solids; but nevertheless an accident common to all the rest follows upon even these geometrical ratios: an infinite possible section of magnitudes is presupposed, which those ratios whose terms are mutually incommensurable somehow involve in actuality too. But the harmonic ratios are all rational, the terms of all are commensurable and are taken from a definite and finite species of plane figures. But infinity of section represents the material, while commensurability or rationality of terms represents the form. Accordingly, as material desires the form, as the rough-hewn stone, of a just magnitude indeed, the form of a human body, so the geometric ratios of figures desire the consonances—not in order to fashion and form those consonances, but because this material squares better with this form, this quantity of stone with this statue, even this ratio of regular solids with this consonance—therefore in order so that they are fashioned and formed more fully, the material by its form, the stone by the chisel into the form of an animate being; but the ratio of the spheres of the figure by its own, *i.e.*, the near and fitting, consonance.

The things which have been said up to now will become clearer from the history of my discoveries. Since I had fallen into this speculation twenty-four years ago, I first inquired whether the single planetary spheres are equal distances apart from one another (for the spheres are apart in Copernicus, and do not touch one another), that is to say, I recognized nothing more beautiful than the ratio of equality. But this ratio is without head or tail: for this material equality furnished no definite number of mobile bodies, no definite magnitude for the intervals. Accordingly, I meditated upon the similarity of the intervals to the spheres, *i.e.*, upon the proportionality. But the same complaint followed. For although to be sure, intervals which were altogether unequal were produced between the spheres, yet they were not unequally equal, as Copernicus wishes, and neither the magnitude of the ratio nor the number of the spheres was given. I passed on to the regular plane figures: [322] intervals were formed from them by the ascription of circles. I came to the five regular solids: here both the number of the bodies and approximately the true magnitude of the intervals was disclosed, in such fashion that I summoned to the perfection of astronomy the discrepancies remaining over and above. Astronomy was perfect these twenty years; and behold! there was still a discrepancy between the intervals and the regular solids, and the reasons for the distribution of unequal eccentricities among the planets were not disclosed. That is to say, in this house the world, I was asking not only why stones of a more elegant form but also what form would fit the stones, in my ignorance that the Sculptor had fashioned them in the very articulate image of an animated body. So, gradually, especially during these last three years, I came to the consonances and abandoned the regular solids in respect to minima, both because the consonances stood on the side of the form which the finishing touch would give, and the regular solids, on that of the material—which in the world is the number of bodies and the rough-hewn amplitude of the intervals—and also because the consonances gave the eccentricities, which the regular solids did not even promise—that is to say, the consonances made the nose, eyes, and remaining limbs a part of the statue, for which the regular solids had prescribed merely the outward magnitude of the rough-hewn mass.

Wherefore, just as neither the bodies of animate beings are made nor blocks of stone are usually made after the pure rule of some geometrical figure, but something is taken away from the outward spherical figure, however elegant it may be (although the just magnitude of the bulk remains), so that the body may be able to get the organs necessary for life, and the stone the image of the animate being; so too as the ratio which the regular solids had been going to prescribe for the planetary spheres is inferior and looks only towards the body and material, it has to yield to the consonances, in so far as that was necessary in order for the consonances to be able to stand closely by and adorn the movement of the globes.

The other branch of the *envoi*, which concerns universal consonances, has a proof closely related to the first. (As a matter of fact, it was in part assumed above, in XVIII, among the Axioms.) For the finishing touch of perfection, as it were, is due rather to that which perfects the world more; and conversely that thing which occupies a second position is to be detracted from, if either is to be detracted from. But the universal harmony of all perfects the world more than the single twin consonances of different neighbouring twos. For harmony is a certain ratio of unity; accordingly the planets are more united, if they all are in concord together in one harmony, than if each two concord separately in two consonances. Wherefore, in the conflict of both, either one of the two single consonances of two planets was due to yield, so that the universal harmonies of all could stand. But the greater consonances, those of the diverging movements, were due to yield rather than the lesser, those of the converging movements. For if the divergent movements diverge, then they look not towards the planets of the given pair but towards other neighbouring planets, and if the converging movements converge, then the movements of one planet are converging toward the movement of the other, conversely: for example, in the pair Jupiter and Mars the aphelial movement of Jupiter verges toward Saturn, the perihelial of Mars towards the Earth: but the perihelial movement of Jupiter verges toward Mars, the aphelial of Mars toward Jupiter. Accordingly the consonance of the converging movements is more proper to Jupiter and Mars; the consonance of the diverging movements is somehow more foreign to Jupiter and Mars. But the ratio of union which brings together neighbouring planets by twos and twos is less disturbed if the consonance which is more foreign and more removed from them should be adjusted than if the private ratio should be, *viz.*, the one which exists between the more neighbouring movements of neighbouring planets. None the less this adjustment was not very great. For the proportionality has been found in which may stand the universal consonances of all the planets may exist (and these in two distinct modes), and in which (with a certain latitude of tuning merely equal to a comma) may also be embraced the single consonances of two neighbouring planets; the consonances of the converging movements in four pairs, perfect, of the aphelial movements in one pair, of the perihelial movements in two pairs, likewise perfect; the consonances of the diverging movements in four pairs, these, however, within the difference of one diesis (the very small interval by which the human voice [323] in figured song nearly always errs; the single consonance of Jupiter and Mars, this between the diesis and the semitone. Accordingly it is apparent that this mutual yielding is everywhere very good.

Accordingly let this do for our *envoi* concerning the work of God the Creator. It now remains that at last, with my eyes and hands removed from the tablet of demonstrations and lifted up towards the heavens, I should pray, devout and supplicating, to the Father of lights: *O Thou Who dost by the light of nature promote in us the desire for the light of grace, that by its means Thou mayest transport us into the light of glory, I give thanks to Thee, O Lord Creator, Who hast delighted me with Thy makings and in the works of Thy hands have I exulted. Behold! now, I have completed the work of my profession, having employed as much power of mind as Thou didst give to me; to the men who are going to read those demonstrations I have made manifest the glory of Thy works, as much of its infinity as the narrows of my intellect could apprehend. My mind has been given over to philosophizing most correctly: if there is anything unworthy of Thy designs brought forth by me—a worm born and nourished in a wallowing place of sins—breathe into me also that which Thou dost wish men to know, that I may make the correction: If I have been allured into rashness by the wonderful beauty of Thy works, or if I have loved my own glory among men, while I am advancing in the work destined for Thy glory, be gentle and merciful and pardon me; and finally deign graciously to effect that these demonstrations give way to Thy glory and the salvation of souls and nowhere be an obstacle to that.*

10. Epilogue Concerning the Sun, by way of Conjecture[1]

From the celestial music to the hearer, from the Muses to Apollo the leader of the Dance, from the six planets revolving and making consonances to the Sun at the centre of all the circuits, immovable in place but rotating into itself. For although the harmony is most absolute between the extreme planetary movements, not with respect to the true speeds through the ether but with respect to the angles which are formed by joining with the centre of the sun the termini of the diurnal arcs of the planetary orbits; while the harmony does not adorn the termini, *i.e.*, the single movements, in so far as they are considered in themselves but only in so far as by being taken together and compared with one another, they become the object of some mind; and although no object is ordained in vain, without the existence of some thing which may be moved by it, while those angles seem to presuppose some action similar to our eyesight or at least to that sense-perception whereby, in Book IV, the sublunary nature perceived the angles of rays formed by the planets on the Earth: still it is not easy for dwellers on the Earth to conjecture what sort of sight is present in the sun, what eyes there are, or what other instinct there is for perceiving those angles

[1]See Kepler's commentary on this epilogue in the *Epitome*, pages 10-11.

even without eyes and for evaluating the harmonies of the movements entering into the antechamber of the mind by whatever doorway, and finally what mind there is in the sun. None the less, however those things may be, this composition of the six primary spheres around the sun, cherishing it with their perpetual revolutions and as it were adoring it (just as, separately, four moons accompany the globe of Jupiter, two Saturn, but a single moon by its circuit encompasses, cherishes, fosters the Earth and us its inhabitants, and ministers to us) and this special business of the harmonies, which is a most clear footprint of the highest providence over solar affairs, now being added to that consideration, [324] wrings from me the following confession: not only does light go out from the sun into the whole world, as from the focus or eye of the world, as life and heat from the heart, as every movement from the King and mover, but conversely also by royal law these returns, so to speak, of every lovely harmony are collected in the sun from every province in the world, nay, the forms of movements by twos flow together and are bound into one harmony by the work of some mind, and are as it were coined money from silver and gold bullion; finally, the curia, palace, and praetorium or throne-room of the whole realm of nature are in the sun, whatsoever chancellors, palatines, prefects the Creator has given to nature: for them, whether created immediately from the beginning or to be transported hither at some time, has He made ready those seats. For even this terrestrial adornment, with respect to its principal part, for quite a long while lacked the contemplators and enjoyers, for whom however it had been appointed; and those seats were empty. Accordingly the reflection struck my mind, what did the ancient Pythagoreans in Aristotle mean, who used to call the centre of the world (which they referred to as the "fire" but understood by that the sun) "the watchtower of Jupiter," Διὸς φυλακήν; what, likewise, was the ancient interpreter pondering in his mind when he rendered the verse of the Psalm as: "He has placed His tabernacle in the sun."

But also I have recently fallen upon the hymn of Proclus the Platonic philosopher (of whom there has been much mention in the preceding books), which was composed to the Sun and filled full with venerable mysteries, if you excise that one κλῦθ (hear me) from it; although the ancient interpreter already cited has explained this to some extent, viz., in invoking the sun, he understands Him Who has placed His tabernacle in the sun. For Proclus lived at a time in which it was a crime, for which the rulers of the world and the people itself inflicted all punishments, to profess Jesus of Nazareth, God Our Savior, and to contemn the gods of the pagan poets (under Constantine, Maxentius, and Julian the Apostate). Accordingly Proclus, who from his Platonic philosophy indeed, by the natural light of the mind, had caught a distant glimpse of the Son of God, that true light which lighteth every man coming into this world, and who already knew that divinity must never be sought with a superstitious mob in sensible things, nevertheless perferred to seem to look for God in the sun rather than in Christ a sensible man, in order that at the same time he might both deceive the pagans by honoring verbally the Titan of the poets and devote himself to his philosophy, by drawing away both the pagans and the Christians from sensible beings, the pagans from the visible sun, the Christians from the Son of Mary, because, trusting too much to the natural light of reason, he spit out the mystery of the Incarnation; and finally that at the same time he might take over from them and adopt into his own philosophy whatever the Christians

242   KEPLER

had which was most divine and especially consonant with Platonic philosophy.[1]
And so the accusation of the teaching of the Gospel concerning Christ is laid
against this hymn of Proclus, in its own matters: let that Titan keep as his
private possessions χρῦσα ἠνία [golden reins] and ταμιεῦνν φαοῦς, μεσσατίην, αἰθέρος
ἔδρην, κοδμοῦ κραδιαῖον ἐριφεγγεᾷ κυκλόν [a treasury of light, a seat at the midpart
of the ether, a radiant circle at the heart of the world], which visible aspect
Copernicus too bestows upon him; let him even keep his παλιννοστοὺς διφρείς
[cyclical chariot-drivings], although according to the ancient Pythagoreans he
does not possess them but in their place τὸ κέντρον, Διὸς φυλακήν [the centre, the
watchtower of Zeus]—which doctrine, misshapen by the forgetfulness of ages,
as by a flood, was not recognized by their follower Proclus; let him also keep
his γενεθλὴν Βλαστησασαν [offspring born] of himself, and whatever else is of
nature; in turn, let the philosophy of Proclus yield to Christian doctrines, [325]
let the sensible sun yield to the Son of Mary, the Son of God, Whom Proclus
addresses under the name of the Titan, ζωαρκεὸς, ὦ ἄνα, πηγῆς αὐτὸς ἔχων κλήδα
[O lord, who dost hold the key of the life-supporting spring], and that πάντα τεῆς
ἐπλήσας ἐλερσινοοῖο προνόιης [thou didst fulfill all things with thy mind-awakening
foresight], and that immense power over the μοιράων [fates], and things which
were read of in no philosophy before the promulgation of the Gospel[2], the
demons dreading him as their threatening scourge, the demons lying in ambush
for souls, ὄφρα ὑφιτενοῦς λαθοῖντο πατρὸς περιφέγγεος αὐλῆς [in order that they
might escape the notice of the light-filled hall of the lofty father]; and who
except the Word of the Father is that εἰκὼν παγγενετᾳο θεοῦ, οὗ φάεντος ἀπ' ἀρρήτου
γενετῆρος παύσατο στοιχεῖων ὀρυμάγδος ἐπ ἀλληλοῖσιν ἰόντων [image of the all-beget-
ting father, upon whose manifestation from an ineffable mother the sin of the elements changing into one another ceased], according to the following: *The Earth
was unwrought and a chaotic mass, and darkness was upon the face of the abyss,
and God divided the light from the darkness, the waters from the waters, the sea from
the dry land;* and: *all things were made by the very Word.* Who except Jesus of
Nazareth the Son of God, ψυχῶν ἀναγωγεύς [the shepherd of souls], to whom ἱκεσιὴ
πολυδάκρυος [the prayer of a tearful suppliant] is to be offered, in order that He
cleanse us from sins and wash us of the filth τῆς γενεθλῆς [of generation]—as if
Proclus acknowledged the fomes of original sin—and guard us from punish-
ment and evil, πρηνύωὴ θόον ὄμμα δικῆς [by making mild the quick eye of justice],
namely, the wrath of the Father? And the other things we read of, which
are as it were taken from the hymn of Zacharias (or, accordingly, was that
hymn a part of the *Metroace?*) Αχλὺν ἀποσκεδάσας ὀλεσίμβροτού ἰολαχεύτόν
[dispersing the poisonous, man-destroying mist], *viz.,* in order that He may
give to souls living in darkness and the shadows of death the φάος ἀγνὸν
[holy light] and ὄλβόν ἀστυφελικτὸν ἀπ' εὐσεβίνέρατείης [unshaken happiness from

---

[1]It was the judgment of the ancients concerning his book *Metroace* that in it he set forth,
not without divine rapture, his universal doctrine concerning God; and by the frequent tears
of the author apparent in it all suspicion was removed from the hearers. None the less this
same man wrote against the Christians eighteen epichiremata, to which John Philoponus opposed himself, reproaching Proclus with ignorance of Greek thought, which none the less he
had undertaken to defend. That is to say, Proclus concealed those things which did not make
for his own philosophy.

[2]Nevertheless in Suidas some similar things are attributed to ancient Orpheus, nearly
equal to Moses, as if his pupil; see too the hymns of Orpheus, on which Proclus wrote commentaries.

lovely piety]; for that is to serve God in holiness and justice all our days. Accordingly, let us separate out these and similar things and restore them to the doctrine of the Catholic Church to which they belong. But let us see what the principal reason is why there has been mention made of the hymn. For this same sun which ὕψοθεν ἀρμνίης ῥύμα πλοῦσιον ἐξοτεύει [sluices the rich flow of harmony from on high]—so too Orpheus κόσμου τὸν ἐναρμόνιον δρόμον ἕλκων [making move the harmonious course of the world]—the same, concerning whose stock Phoebus about to rise κιθαρῇ ὑπὸ θέσκελα μελπῶν εὐνάξει μεγὰ κῦμα βαρυφλσισβοῖο γενεθλῆς [sings marvellous things on his lyre and lulls to sleep the heavy-sounding surge of generation] and in whose dance Paean is the partner, πλήσας ἀρμονίης πανατήμονος εὕρεα κόσμν [striking the wide sweep of innocent harmony]—him, I say, does Proclus at once salute in the first verse of the hymn as πῦρος νοεροῦ βασιλέα [king of intellectual fire]. By that commencement, at the same time, he indicates what the Pythagoreans understood by the word of fire (so that it is surprising that the pupil should disagree with the masters in the position of the centre) and at the same time he transfers his whole hymn from the body of the sun and its quality and light, which are sensibles, to the intelligibles, and he has assigned to that πῦρ νοερὸς [intellectual fire] of his—perhaps the artisan fire of the Stoics—to that created God of Plato, that chief or self-ruling mind, a royal throne in the solar body, confounding into one the creature and Him through Whom all things have been created. But we Christians, who have been taught to make better distinctions, know that this eternal and uncreated "Word," Which was "with God" and Which is contained by no abode, although He is within all things, excluded by none, although He is outside of all things, took up into unity of person flesh out of the womb of the most glorious Virgin Mary, and, when the ministry of His flesh was finished, occupied as His royal abode the heavens, wherein by a certain excellence over and above the other parts of the world, viz., through His glory and majesty, His celestial Father too is recognized to dwell, and has also promised to His faithful, mansions in that house of His Father: as for the remainder concerning that abode, we believe it superfluous to inquire into it too curiously or to forbid the senses or natural reasons to investigate that which the eye has not seen nor the ear heard and into which the heart of man has not ascended; but we duly subordinate the created mind—of whatsoever excellence it may be—to its Creator, and we introduce neither God-intelligences with Aristotle and the pagan philosophers nor armies of innumerable planetary spirits with the Magi, nor do we propose that they are either to be adored or summoned to intercourse with us by theurgic superstitions, for we have a careful fear of that; but we freely inquire by natural reasons what sort of thing each mind is, especially if in the heart of the world [326] there is any mind bound rather closely to the nature of things and performing the function of the soul of the world—or if also some intelligent creatures, of a nature different from human perchance do inhabit or will inhabit the globe thus animated (see my book on the New Star, Chapter 24, "On the Soul of the World and Some of Its Functions"). But if it is permissible, using the thread of analogy as a guide, to traverse the labyrinths of the mysteries of nature, not ineptly, I think, will someone have argued as follows: The relation of the six spheres to their common centre, thereby the centre of the whole world, is also the same as that of διανοία [discussive intellection] to νοῦς [intuitive intellection], according as these faculties are distinguished by Aristotle, Plato, Proclus, and the rest; and the relation of the single planets' revolutions in place around the sun to the ἀμετάθεδον [unvarying] rotation of the sun in the central space of the whole system (concerning which the sun-spots are evidence; this has been demonstrated in the *Commentaries on the Movement of Mars*) is the same as the relation of τὸ διανοητικὸν to τὸ νοερὸν, that of the manifold discourses of ratiocination to the most simple intellection of the mind. For as the sun rotating into itself moves all the planets by means of the form emitted from itself, so too—as the philosophers teach—mind, by understanding itself and in itself all things, stirs up ratiocinations, and by dispersing and unrolling its simplicity into them, makes everything to be understood. And the movements of the planets around the sun at their centre and the discourses of ratiocinations are so interwoven and bound together that, unless the Earth, our domicile, measured out the annual circle, midway between the other spheres—changing from place to place, from station to station—never would human ratiocination have worked its way to the true intervals of the planets and to the other things dependent from them, never would it have constituted astronomy. (See the *Optical Part of Astronomy*, Chapter 9.)

On the other hand, in a beautiful correspondence, simplicity of intellection follows upon the stillness of the sun at the centre of the world, in that hitherto we have always worked under the assumption that those solar harmonies of movements are defined neither by the diversity of regions nor by the amplitude of the expanses of the world. As a matter of fact, if any mind observes from the sun those harmonies, that mind is without the assistance afforded by the movement and diverse stations of his abode, by means of which it may string together ratiocinations and discourse necessary for measuring out the planetary intervals. Accordingly, it compares the diurnal movements of each planet, not as they are in their own orbits but as they pass through the angles at the centre of the sun. And so if it has knowledge of the magnitude of the spheres, this knowledge must be present in it *a priori*, without any toil of ratiocination: but to what extent that is true of human minds and of sublunary nature has been made clear above, from Plato and Proclus.

Under these circumstances, it will not have been surprising if anyone who has been thoroughly warmed by taking a fairly liberal draft from that bowl of Pythagoras which Proclus gives to drink from in the very first verse of the hymn, and who has been made drowsy by the very sweet harmony of the dance of the planets begins to dream (by telling a story he may imitate Plato's Atlantis and, by dreaming, Cicero's Scipio): throughout the remaining globes, which follow after from place to place, there have been disseminated discursive or ratiocinative faculties, whereof that one ought assuredly to be judged the most excellent and absolute which is in the middle position among those globes, *viz.*, in man's earth, while there dwells in the sun simple intellect, πῦρ νοερὸν, or νοῦς, the source, whatsoever it may be, of every harmony.

For if it was Tycho Brahe's opinion concerning that bare wilderness of globes that it does not exist fruitlessly in the world but is filled with inhabitants: with how much greater probability shall we make a conjecture as to God's works and designs even for the other globes, from that variety which we discern in this globe of the Earth. For He Who created the species which should inhabit the waters, beneath which however there is no room for the air [327] which living things draw in; Who sent birds supported on wings into the wilderness of the air; Who gave white bears and white wolves to the snowy regions of the North, and as food for the bears the whale, and for the wolves, birds' eggs; Who gave lions to the deserts of burning Libya and camels to the wide-spread plains of Syria, and to the lions an endurance of hunger, and to the camels an endurance of thirst: did He use up every art in the globe of the Earth so that He was unable, every goodness so that he did not wish, to adorn the other globes too with their fitting creatures, as either the long or short revolutions, or the nearness or removal of the sun, or the variety of eccentricities or the shine or darkness of the bodies, or the properties of the figures wherewith any region is supported persuaded?

Behold, as the generations of animals in this terrestrial globe have an image of the male in the dodecahedron, of the female in the icosahedron—whereof the dodecahedron rests on the terrestrial sphere from the outside and the icosahedron from the inside: what will we suppose the remaining globes to have, from the remaining figures? For whose good do four moons encircle Jupiter, two Saturn, as does this our moon this our domicile? But in the same way we shall ratiocinate concerning the globe of the sun also, and we shall as it were incorporate conjectures drawn from the harmonies, et cetera—which are weighty of themselves—with other conjectures which are more on the side of the bodily, more suited for the apprehension of the vulgar. Is that globe empty and the others full, if everything else is in due correspondence? If as the Earth breathes forth clouds, so the sun black smoke? If as the Earth is moistened and grows under showers, so the sun shines with those combusted spots, while clear flamelets sparkle in its all fiery body. For whose use is all this equipment, if the globe is empty? Indeed, do not the senses themselves cry out that fiery bodies dwell here which are receptive of simple intellects, and that truly the sun is, if not the king, at least the queen πῦρος νοεροῦ [of intellectual fire]?

Purposely I break off the dream and the very vast speculation, merely crying out with the royal Psalmist: *Great is our Lord and great His virtue and of His wisdom there is no number: praise Him, ye heavens, praise Him, ye sun, moon, and planets, use every sense for perceiving, every tongue for declaring your Creator. Praise Him, ye celestial harmonies, praise Him, ye judges of the harmonies uncovered* (and you before all, old happy Mastlin, for you used to animate these cares with words of hope): *and thou my soul, praise the Lord thy Creator, as long as I shall be: for out of Him and through Him and in Him are all things,* καὶ τὰ αἰσθητὰ καὶ τὰ νοερὰ *[both the sensible and the intelligible]; for both whose whereof we are utterly ignorant and those which we know are the least part of them; because there is still more beyond. To Him be praise, honour, and glory, world without end. Amen.*

THE END

This work was completed on the 17th or 27th day of May, 1618; but Book v was reread (while the type was being set) on the 9th or 19th of February, 1619. At Linz, the capital of Austria—above the Enns.

GREAT BOOKS IN PHILOSOPHY PAPERBACK SERIES

ESTHETICS

❑ Aristotle—*The Poetics*
❑ Aristotle—*Treatise on Rhetoric*

ETHICS

❑ Aristotle—*The Nicomachean Ethics*
❑ Marcus Aurelius—*Meditations*
❑ Jeremy Bentham—*The Principles of Morals and Legislation*
❑ John Dewey—*The Moral Writings of John Dewey, Revised Edition*
  (edited by James Gouinlock)
❑ Epictetus—*Enchiridion*
❑ Immanuel Kant—*Fundamental Principles of the Metaphysic of Morals*
❑ John Stuart Mill—*Utilitarianism*
❑ George Edward Moore—*Principia Ethica*
❑ Friedrich Nietzsche—*Beyond Good and Evil*
❑ Plato—*Protagoras, Philebus,* and *Gorgias*
❑ Bertrand Russell—*Bertrand Russell On Ethics, Sex, and Marriage*
  (edited by Al Seckel)
❑ Arthur Schopenhauer—*The Wisdom of Life* and *Counsels and Maxims*
❑ Benedict de Spinoza—*Ethics* and *The Improvement of the Understanding*

METAPHYSICS/EPISTEMOLOGY

❑ Aristotle—*De Anima*
❑ Aristotle—*The Metaphysics*
❑ Francis Bacon—*Essays*
❑ George Berkeley—*Three Dialogues Between Hylas and Philonous*
❑ W. K. Clifford—*The Ethics of Belief and Other Essays*
  (introduction by Timothy J. Madigan)
❑ René Descartes—*Discourse on Method* and *The Meditations*
❑ John Dewey—*How We Think*
❑ John Dewey—*The Influence of Darwin on Philosophy and Other Essays*
❑ Epicurus—*The Essential Epicurus: Letters, Principal Doctrines, Vatican Sayings,
  and Fragments* (translated, and with an introduction, by Eugene O'Connor)
❑ Sidney Hook—*The Quest for Being*
❑ David Hume—*An Enquiry Concerning Human Understanding*
❑ David Hume—*Treatise of Human Nature*
❑ William James—*The Meaning of Truth*
❑ William James—*Pragmatism*
❑ Immanuel Kant—*The Critique of Judgment*
❑ Immanuel Kant—*Critique of Practical Reason*
❑ Immanuel Kant—*Critique of Pure Reason*
❑ Gottfried Wilhelm Leibniz—*Discourse on Metaphysics* and *the Monadology*
❑ John Locke—*An Essay Concerning Human Understanding*
❑ Charles S. Peirce—*The Essential Writings*
  (edited by Edward C. Moore, preface by Richard Robin)
❑ Plato—*The Euthyphro, Apology, Crito,* and *Phaedo*
❑ Plato—*Lysis, Phaedrus,* and *Symposium*
❑ Bertrand Russell—*The Problems of Philosophy*
❑ George Santayana—*The Life of Reason*
❑ Sextus Empiricus—*Outlines of Pyrrhonism*

PHILOSOPHY OF RELIGION

- ❏ Marcus Tullius Cicero—*The Nature of the Gods* and *On Divination*
- ❏ Ludwig Feuerbach—*The Essence of Christianity*
- ❏ David Hume—*Dialogues Concerning Natural Religion*
- ❏ John Locke—*A Letter Concerning Toleration*
- ❏ Lucretius—*On the Nature of Things*
- ❏ John Stuart Mill—*Three Essays on Religion*
- ❏ Friedrich Nietzsche—*The Antichrist*
- ❏ Thomas Paine—*The Age of Reason*
- ❏ Bertrand Russell—*Bertrand Russell On God and Religion* (edited by Al Seckel)

SOCIAL AND POLITICAL PHILOSOPHY

- ❏ Aristotle—*The Politics*
- ❏ Mikhail Bakunin—*The Basic Bakunin: Writings, 1869–1871* (translated and edited by Robert M. Cutler)
- ❏ Edmund Burke—*Reflections on the Revolution in France*
- ❏ John Dewey—*Freedom and Culture*
- ❏ John Dewey—*Individualism Old and New*
- ❏ John Dewey—*Liberalism and Social Action*
- ❏ G. W. F. Hegel—*The Philosophy of History*
- ❏ G. W. F. Hegel—*Philosophy of Right*
- ❏ Thomas Hobbes—*The Leviathan*
- ❏ Sidney Hook—*Paradoxes of Freedom*
- ❏ Sidney Hook—*Reason, Social Myths, and Democracy*
- ❏ John Locke—*Second Treatise on Civil Government*
- ❏ Niccolo Machiavelli—*The Prince*
- ❏ Karl Marx (with Friedrich Engels)—*The German Ideology*, including *Theses on Feuerbach* and *Introduction to the Critique of Political Economy*
- ❏ Karl Marx—*The Poverty of Philosophy*
- ❏ Karl Marx/Friedrich Engels—*The Economic and Philosophic Manuscripts of 1844* and *The Communist Manifesto*
- ❏ John Stuart Mill—*Considerations on Representative Government*
- ❏ John Stuart Mill—*On Liberty*
- ❏ John Stuart Mill—*On Socialism*
- ❏ John Stuart Mill—*The Subjection of Women*
- ❏ Friedrich Nietzsche—*Thus Spake Zarathustra*
- ❏ Thomas Paine—*Common Sense*
- ❏ Thomas Paine—*Rights of Man*
- ❏ Plato—*The Republic*
- ❏ Jean-Jacques Rousseau—*The Social Contract*
- ❏ Mary Wollstonecraft—*A Vindication of the Rights of Men*
- ❏ Mary Wollstonecraft—*A Vindication of the Rights of Women*

GREAT MINDS PAPERBACK SERIES

CRITICAL ESSAYS

- ❏ Desiderius Erasmus—*The Praise of Folly*
- ❏ Jonathan Swift—*A Modest Proposal and Other Satires* (with an introduction by George R. Levine)
- ❏ H. G. Wells—*The Conquest of Time* (with an introduction by Martin Gardner)

ECONOMICS

- ❏ Charlotte Perkins Gilman—*Women and Economics: A Study of the Economic Relation between Women and Men*
- ❏ John Maynard Keynes—*The General Theory of Employment, Interest, and Money*

❑ John Maynard Keynes—*A Tract on Monetary Reform*
❑ Thomas R. Malthus—*An Essay on the Principle of Population*
❑ Alfred Marshall—*Principles of Economics*
❑ Karl Marx—*Theories of Surplus Value*
❑ David Ricardo—*Principles of Political Economy and Taxation*
❑ Adam Smith—*Wealth of Nations*
❑ Thorstein Veblen—*Theory of the Leisure Class*

HISTORY

❑ Edward Gibbon—*On Christianity*
❑ Alexander Hamilton, John Jay, and James Madison—*The Federalist*
❑ Herodotus—*The History*
❑ Thucydides—*History of the Peloponnesian War*
❑ Andrew D. White—*A History of the Warfare of Science with Theology in Christendom*

LAW

❑ John Austin—*The Province of Jurisprudence Determined*

PSYCHOLOGY

❑ Sigmund Freud—*Totem and Taboo*

RELIGION

❑ Thomas Henry Huxley—*Agnosticism and Christianity and Other Essays*
❑ Ernest Renan—*The Life of Jesus*
❑ Upton Sinclair—*The Profits of Religion*
❑ Elizabeth Cady Stanton—*The Woman's Bible*
❑ Voltaire—*A Treatise on Toleration and Other Essays*

SCIENCE

❑ Nicolaus Copernicus—*On the Revolutions of Heavenly Spheres*
❑ Charles Darwin—*The Autobiography of Charles Darwin*
❑ Charles Darwin—*The Descent of Man*
❑ Charles Darwin—*The Origin of Species*
❑ Charles Darwin—*The Voyage of the* Beagle
❑ Albert Einstein—*Relativity*
❑ Michael Faraday—*The Forces of Matter*
❑ Galileo Galilei—*Dialogues Concerning Two New Sciences*
❑ Ernst Haeckel—*The Riddle of the Universe*
❑ William Harvey—*On the Motion of the Heart and Blood in Animals*
❑ Werner Heisenberg—*Physics and Philosophy: The Revolution in Modern Science*
    (introduction by F. S. C. Northrop)
❑ Julian Huxley—*Evolutionary Humanism*
❑ Edward Jenner—*Vaccination against Smallpox*
❑ Johannes Kepler—*Epitome of Copernican Astronomy and Harmonies of the World*
❑ Isaac Newton—*The Principia*
❑ Louis Pasteur and Joseph Lister—*Germ Theory and Its Application to Medicine
    and On the Antiseptic Principle of the Practice of Surgery*
❑ Alfred Russel Wallace—*Island Life*

SOCIOLOGY

❑ Emile Durkheim—*Ethics and the Sociology of Morals*
    (translated with an introduction by Robert T. Hall)